Creative
Radio
Production

Bruce H. Siegel

Department of Communication Studies
State University of New York

Focal Press
Boston London

Focal Press is an imprint of Butterworth–Heinemann.

Recognizing the importance of preserving what has been written, it is the policy of Butterworth–Heinemann to have the books it publishes printed on acid-free paper, and we exert our best efforts to that end.

Library of Congress Cataloging-in-Publication Data

Siegel, Bruce H.
 Creative radio production / by Bruce H. Siegel.
  p.   cm.
 Includes bibliographical references (p. 280) and index.
 ISBN 0-240-80070-2
 1. Radio—Production and direction. 2. Radio—Apparatus and supplies. 3. Sound—Recording and reproducing. I. Title.
PN1991.75.S54   1992
791.44′0232—dc20   91-30081
                                                              CIP

British Library Cataloguing in Publication Data

Siegel, Bruce H.
 Creative radio production.
 I. Title
 791.440232

ISBN 0-240-80070-2

Butterworth–Heinemann
80 Montvale Avenue
Stoneham, MA 02180

10 9 8 7 6 5 4 3 2 1

Printed in the United States of America

This book is fondly dedicated

to my parents, James and Rebecca Siegel, who probably aren't surprised

to Charles Abraham Rovner, who, I hope, knows

to Mrs. Dolores R. Keefe and Miss Lenore Mooney for showing me how to make music

to my son, Joshua, who, I hope, will always find joy and wonder in sound

and to my wife, Anne, with whom everything is possible, and without whom nothing matters.

Contents

Preface

When I was six, my Dad showed me a card trick that I remember to this day. He riffled through the deck instructing me to say "stop" whenever I felt like it, and when I said "stop," he showed me the card I'd stopped at and then had me bury it somewhere in the deck. Then he looked through the cards and pulled one out saying, "Is this your card?" It wasn't. He put the card face down on the table and tried again.

"How about this one?"

"Nope."

Appearing exasperated (an Oscar-winning performance if there ever was one), he put this second one facedown on the table alongside his first error.

"Give me one more chance."

More fumbling with the deck, another card chosen, another failure.

There were now three cards facedown on the table, none of them my chosen card. I smiled smugly. I was only six, and I had him.

"Hmm," he said thoughtfully. He turned over one of the three tabled cards. "You're sure this isn't your card?"

"Positive."

He turned over the second card. "And this one isn't it either?"

"Uh-uh."

"Well what was your card?"

I told him.

"That's what I thought." And he turned over the last card, and lo, it was the card I'd chosen. At that instant, I was stunned. I'd witnessed a miracle. My Dad had single-handedly put the laws of time, space, and nature on hold and had made a wrong card turn into the right card. It was wonderful—not just wonderful, but *WONDER-full*. Full of wonder. Inspiring amazement, awe. Magic in its purest form.

Wonder is a central fixture of childhood. Kids live in a world of imagination. But something happens around age 9 or 10. Socialization. Suddenly, the child's dependence on imagination all but disappears. Magic isn't cool anymore, so the imagination gets stuffed into a closet in the back of the psyche. Occasionally, the closet door cracks open, like when we hear something go bump in the night, only to be slammed shut again when the light goes on, illuminating the family cat and an upended lamp.

For all our sophistication, there's a large part of each of us that still loves to experience wonder. That's why we love to witness great beauty—natural or manmade. That's why we secretly hope they never catch the Loch Ness Monster. That's why magicians like Sigfried and Roy play to packed houses in Las Vegas.

That's why I've written this book.

To show you how to use a simple commodity like sound to make magic. To create illusion. To inspire, on a small scale, wonder. You are a creative, imaginative being. I'm convinced that there's a gene on a chromosome somewhere that makes all of us, simply by dint of being members of the *Homo sapiens* club, inherently creative. Some people have a lot of creativity. We call them artists. Or geniuses (the most overworked and ill-used word in the language). The rest of us have to make do with what creativity we have, but we all have *some*. The problem is that you may not realize that you are creative by nature. All you need to do is open that closet and express what emerges through a medium of some sort like color, or words, or movement, or a lens.

Or sound.

Sound is my medium. With it, you and I can paint pictures as colorful as those of any artist. We can make an audience smile. Or shiver. Or cheer. We can sell a product or an idea. We can make people feel comfortable, uneasy, or ashamed. We can do all this and more—all without being seen. My medium, sound, is the theater of the imagination; and perhaps most important, it's a medium that anyone can master.

Anyone.

What say we open your closet?

Bruce Siegel
State University of New York

Acknowledgments

If I were to categorize the people who've been instrumental in getting this book into your hands, they'd fall under two headings. First, those who helped bring this project to fruition. People like Phil Sutherland, Evelyn Laiacona, and Mary Cervantes at Focal Press, and Dolores Wolfe at York Production Services. The fact that they've been able to weather my ideas, complaints, and whims for so many months with good humor, patience, and mental and emotional stability says a lot. Thanks go out to Tom Soccocio, Jr., producer par excellence at my alma mater, 62WHEN radio in Syracuse, New York, for his assistance and cheery demeanor. Being an audio person, visual stuff gives me fits, so I made sure that this part of the book was in good hands. Gloria Lerman Taylor, long-time friend, and aficionado of good card tricks was the talent behind the camera. And Mark R. Taylor, spouse of Gloria, artist, musician, all-around good guy, and the only person I know who loves boxing and music as much as I do, was the illustrator.

The second group of people I have to thank are those who taught me what I know: teachers, students, musicians, producers. All those whose paths I've been lucky enough to cross, and to whose influence this book bears witness. The Talmud says that it is incumbent upon every person to do his or her teachers honor. For all my teachers, past, present, and future, this book is my attempt to do just that.

About the Tape

Let's face it! It's pretty difficult (though not impossible) to learn radio production from a book—any book. Trying to adequately describe audio phenomena in a print format is like trying to describe the concept of color to someone who's been blind from birth. To help you (and your teacher, if you're learning production in a class), this book comes with a companion cassette so that you can *hear* what I'm talking about, and not have to be content with only the written word. If you're using the book without the tape, don't worry. The text is designed to be complete in and of itself. It's just that the tape can help smooth out some of the bumps you may encounter along the way.

The icon ⊟ in the text indicates that there's a cut on the tape illustrating the concept or technique you've just learned.

The book includes as Appendix 2, the Table of Contents to the Cassette Tape.

1

□ □ □
□ □ □
□ □ □

Fundamentals of Sound

Before working in any artistic medium, you must be well acquainted with the qualities and quirks of that medium. Whether you are working with color or clay, fabric or film, stone or sound, you must have a basic idea of what the medium is and does before learning what you can do with it. That being said, let me assure you that if you're one of the legion of high school physics flunkers (don't even mention *college* physics), be of stout heart: many of the best audio people I know can't even *spell* physics, much less claim proficiency in it. Needless to say, I'm not here to make you into a scientist or an acoustician or an electronics technician: I'm here to show you how anyone with a modicum of creativity and a basic knowledge of studio equipment can turn out high-quality radio production. Nonetheless, if you want to play with sound, you have to understand sound.

Do you know why good (or even middling) magicians can fool you? They know that the solutions to most illusions are quite simple. You, however, assume that because you've been momentarily surprised by their legerdemain, and because you're an intelligent person, the solution to the trick must be complicated. Exactly the opposite is true! You're looking up in the stratosphere for an answer that's down by your knees. The result: You can't figure it out!

The physics you need to know in order to function adequately in an audio studio is exactly the same—simple. If you're technically minded, there are a number of really fine books on the market chock full of graphs, charts, diagrams, and the like. It's all good information, but not absolutely necessary for our purpose here.

The Nature of Sound

Unlike other artistic media, sound can be perceived only *aurally* (through the ears). It's true that if you crank up your amplifier, and rattle your walls, you will be able to feel the very real pressure of the sound, especially of the bass frequencies. For the most part, though, any concept of sound we may have comes through our ears.

In order for sound to exist, three elements *must be* present:

1. a vibration
2. a medium
3. a receiver

A *vibration* is described as a *periodic movement*—that is, a movement that repeats at *regular intervals of time*. Halley's Comet is periodic. So are the swallows' return to Capistrano and the metronomic stride of a marathon runner. All these activities repeat over the course of various intervals of time. Halley's Comet repeats its action every 76 years, the swallows return once a year, and the runner may churn out two or more strides every second. A pendulum is a perfect example of periodic motion. If you pull it off to one side and release it, it swings back and forth at fairly regular intervals until gravity and air friction gradually slow it down and return it to a position of equilibrium—vertical. Scientists doing the early research into sound needed somehow to create a pictorial representation of this sort of activity. Obviously, because they couldn't see the subject of their experiments, a graphic substitute was in order, and the sine wave did the trick.

Try this: Stand at the left side of a fairly long blackboard. With a piece of chalk in hand, repeatedly move your arm up and down steadily as you slowly walk to the right side of the blackboard. Even though the actual movement of your arm was vertical, the resulting design on the blackboard should approximate a sine wave (see Figure 1.1). A lot of people think that sound looks like this wave. Sound, in fact, does not look like anything. It can't be seen. The sine wave is merely a symbol, a means by which we can understand and analyze a nonvisual entity.

If you look at the sine wave, its repetitiveness is obvious. It undulates regularly up and down. If we place the wave on a horizontal axis, its individual cycles become apparent. The wave in Figure 1.1 (b) begins on the central axis, rises to a peak, curves back down to the axis, crosses the axis, descends a distance equal to the height of the peak, curves back up to the axis, and repeats the sequence. This pattern represents one *cycle* of our vibration. The cycle can begin anywhere on the curve, not just on the axis. The distance from one peak to the next peak is also a complete cycle, as is the distance between consecutive troughs. These lengths are all the same, and all represent one complete cycle. So much for vibration. I told you it was simple.

The next element in the chain of sound transmittance is the *medium*, something for the energy from the vibration to pass through. One of the bizarre moments in the movie *2001: A Space Odyssey* (and it had a lot of them), occurred when an astronaut, stranded outside his capsule by a particularly nasty computer, had to detonate the explosive bolts on a hatchway in order to reenter the vehicle. What was eerie, aside from the premise of the whole thing, was

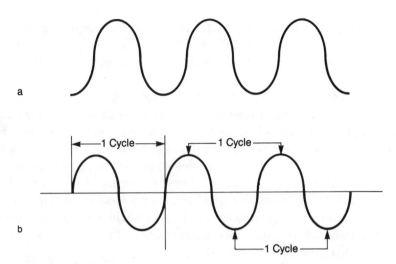

1.1 (a) A sinusoidal (sine) wave shows a graphic representation of periodic motion. (b) A cycle can begin anywhere on the sine wave.

that the explosion, which blew out the hatch and sucked the astronaut into the capsule, was totally silent. In order for the explosion to be heard, it would've had to transmit its vibration through some sort of medium. Because the explosion occurred in space, in a vacuum, there was no medium (such as air) through which the energy could travel and, consequently, no sound. "Come off it," I hear you say. "If I were in outer space, and a stick of dynamite went off next to my head, I wouldn't hear anything?" Right. You'd feel the force of the explosion, but there would be no bang.

Have you ever tried to talk to somebody underwater? You could hear the person, but the sound was probably unintelligible. The reason, that sound is garbled underwater but is clearer in the air has to do with the medium's density and its elasticity. The *elasticity* of a medium is its ability to recover its original form after a force has momentarily pushed or pulled it out of shape. Put another way, elasticity is the force within an object that causes that object to *resist* being bent, twisted, or otherwise pushed out of shape. It sounds strange but the more elastic something is, the *less* flexible it is. A taut rubber band, as you might suspect, is a perfect example. If you push or pull it, and release it, it quickly resumes its original shape. I remember years ago watching a teacher work with a deaf child. She pressed an inflated balloon against the side of her mouth as she spoke, and the child, with hands and face pressed against the other side of the balloon, was able to easily feel the vibration. It's difficult to envision air as having this same elasticity, but it does. The more elastic a medium is, the faster a vibration will pass through it.

Don't equate elasticity with flexibility. If anything, you're better off associating elasticity with *rigidity*. Everything in nature has some degree of elasticity, no matter how hard or inflexible it is. A colleague of mine explains elasticity by saying that the greater the force necessary to bend an object and keep it bent, the greater its elasticity.

So how come your attempt to talk underwater was such a flop? Doesn't water keep its shape more than air does, so isn't water more elastic than air? This is absolutely true, but for a medium to transmit sound *accurately*, you have to consider not only its elasticity, but also its *density*. If a medium is too *dense*—that is, if its atoms and molecules are packed too tightly together—the vibration will indeed pass through it, but it will emerge distorted: the medium will alter the vibration, so that what comes out is very different from what went in. Also, as you might guess, the denser a medium is, the more slowly the vibration will pass through it, not unlike trying to drive across Manhattan at rush hour. Keep these points in mind:

1. The denser the medium, the slower the sound travels.
2. The more elastic the medium, the faster the sound travels.
3. Frequently (though not always), the denser the medium is, the more elastic it is. An exception to this might be liquid mercury, which, although very dense, is nonetheless quite inelastic, that is, it has little resistance to being pushed into a new shape (assuming you can catch it first).

Sound travels at 1,087 feet per second (741 miles per hour) at sea level with the temperature at freezing (32 degrees Fahrenheit, 0 degrees Celsius). As the temperature rises, the air becomes less dense (because the heated air molecules spread out), and the speed of sound goes up 1.1 feet per second (ft/sec) for each additional degree Fahrenheit. If you're at sea level, and the temperature is 65 degrees Fahrenheit (F), the sound would travel 1,123.3 ft/sec. That is, 65° F is 33 degrees above freezing, so

$$1,087 + (1.1 \times 33) = 1,087 + 36.3 = 1,123.3 \text{ ft/sec}$$

Frequently, I use 1,100 feet per second for rough computations because it's easier to juggle than 1,087. Besides, my studio isn't at sea level, and it's certainly warmer than 32. Sound will travel faster in jungle air (which is denser but considerably more elastic) than on top of a mountain (where the air is rarefied and less elastic). Remember, increased density interferes with a vibration's speed; however, if the increase in density is accompanied by an increase in elasticity, the elasticity will accelerate the sound at a greater rate than the density will slow it down.

Here's an example of how this principle works: A toy came on the market a number of years ago, consisting of five steel balls hanging on strings in a small wooden frame. The balls were arranged in a tight, straight line, and when one ball was pulled to the side and released, it smacked the row, knocking the ball on the other end off and up in an arc. This ball, in turn, fell back, hit the stack, causing the first ball to pop up again, and so on and so on. It was a great gadget to keep an executive busy between coffee breaks. Anyway, picture two of these devices, but put a small gap (let's say a sixteenth of an inch) between the balls of one of them. Pull the leftmost ball on both toys the same distance to the side and release them simultaneously. The one with the space between the balls will take much longer to transmit the energy from the initial impact through the line to the end ball. Not only that, but the energy will be dissipated much more rapidly, and the balls will stop swinging much sooner than with the other toy. The reason, believe it or not, is because steel is more elastic than air, so that the toy with the solid steel path will transmit the vibration much more rapidly. Think about it. Push the air, and it resumes its former shape fairly slowly. Slam a steel beam with a hammer, and it resumes its original shape instantly. That's what elasticity is all about—how fast the substance bounces back (or, if you prefer, how much force would be required to bend it). Sound, which travels at about 741 miles per hour through the air, travels at more than 11,000 miles per hour through a steel beam.

The third element in the sound chain is the *receiver*. Remember the old conundrum: If a tree falls in the woods, and there's no one around to hear it, does it make a sound? Technically, no. As a vibration travels through a medium, the only way it can be perceived as sound is by striking an eardrum, or some similar receptor. The *eardrum*, a thin elastic membrane, is then set to vibrating; and this vibration, in turn, is passed on to the brain, which then retranslates the vibration back into what we perceive as sound. These steps seem simple, but in fact, the transformation from vibration to sound within the human ear is wonderfully complex.

In a nutshell, the vibration passes through three chambers: the outer, middle, and inner ears (see Figure 1.2). The outer ear consists basically of the *pinna* (the thing hanging on the side of your head) and the auditory canal, a narrow passageway, (into which only your elbow is supposed to be inserted, right?) which focuses the vibration (like a megaphone in reverse) toward the eardrum. As the vibration strikes the eardrum, causing it, in turn, to vibrate, the eardrum passes the vibration into the middle ear, an area containing three small connected bones (the smallest bones in your body), picturesquely named the *hammer*, the *anvil*, and the *stirrup*. The vibration is passed along this tiny chain of bones to the inner ear, which consists of a small organ that resembles a snail that's undergone some kind of mutation. The largest segment of this organ is called the *cochlea*, and in it we find auditory fluid and auditory nerve endings. Finally, the vibration is passed through the auditory fluid to the auditory nerve, which then carries the vibration along to the brain.

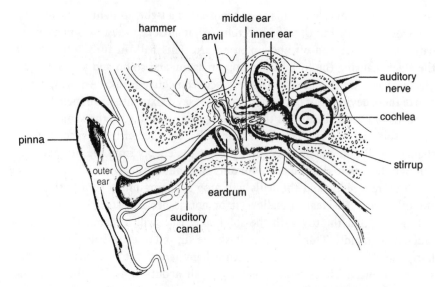

1.2 Cross-section of the human ear. (From AUDIO IN MEDIA Third/Second Editions, by Stanley R. Alten © 1990, 1986 by Wadsworth, Inc. Reprinted by permission of the publisher.)

This process gets the vibration through the labyrinth of your ear to your brain. This we know, but how the brain reprocesses the vibration into the sound you perceive is a mystery—for the time being, anyway. We don't completely understand how we hear sound, and sound, by definition, must be heard. Without a receiver of some sort, the tree falling in the woods produces a vibration—but technically no sound.

Sound Waves

Even though we can't see sound waves, it's easy to visualize what they are and how they act because they're analogous in many respects to water waves. When a stone is tossed into a pond, the resultant disturbance parallels a sound source sending a vibration through the air. If we view the surface of the pond at eye level, we get a picture of the wave(s) moving away from the source of the disturbance. The crest of the wave is higher than the surface of the pond because the stone entering the water displaces the water immediately adjacent to it, forcing it outward, thus creating a bulge on the surface.

Place a piece of paper on a table, and position your hands on opposite ends of the paper. When you move your hands together, the paper bulges upward. The force applied by your hands causes the paper to become compressed, and, having no other way to relieve the pressure, it bulges upward. The wave on

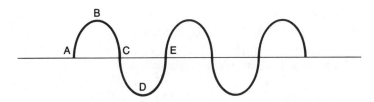

1.3 Segmented sine wave.

the pond is exactly the same. It's actually the compression of a small piece of the pond's surface and the water beneath it. Immediately behind the wave is a trough, a small hollow in the pond's surface. This isn't surprising: the additional water needed to make the crest higher than the surface had to come from somewhere, and, as you might suspect, the height of the crest and the depth of the trough are identical.

If the wave is an area of *compression* (greater pressure), the trough is an area of *rarefaction* (less pressure). The sine wave, which we're using to represent sound, is exactly the same (see Figure 1.3). The segment of the wave from A to C represents the compression of the air as the vibration travels through it; and the segment from C to E is the rarefaction. B is the crest (the greatest pressure), and D is the trough (the least pressure).

It's time to debunk a common misconception. A wave, whether in water or air or wherever, does not really travel through the medium. What actually moves is the *vibration*, the energy if you will. If you watch a sea bird bobbing on the ocean, you'll observe that when a wave reaches it, the bird appears to move *vertically*, not horizontally with the flow of the wave. That's because the water under the bird, despite appearances, is moving up and down, not sideways. It's the *energy* from the initial disturbance that is moving laterally.

There are two types of waves:

1. *Transverse waves*, in which the movement of the wave is perpendicular to the direction of the force that generated it—When we dropped the stone into the pond, the movement of the stone was vertical; however, the force it generated moved out horizontally.
2. *Longitudinal waves*, in which the generating force and the wave travel in the same direction—The toy (with five balls) mentioned previously is an example of this type of wave.

Characteristics of Sound

Despite the staggering variety of sounds present in our lives, there are actually only four characteristics that all sounds have and that differentiate one sound from another:

1. *frequency* (pitch)
2. *amplitude* (loudness)
3. *timbre* (harmonic structure)
4. *envelope* (attack–decay–sustain–release times)

Frequency

Frequency is the number of cycles a vibrating body completes in a given interval of time. On the sine wave in Figure 1.3, the segment from A to E represents a single cycle, after which the pattern repeats itself. Pull a pendulum off to the left and release it. It swings to the right and swings back to the left—one complete cycle—and it continues this cyclical pattern until gravity and air friction return it to a state of vertical equilibrium. On the piano, if we strike the A above middle C and watch the string vibrate, we see a blur because the string is completing 440 pendulumlike cycles every second. The frequency of this sound is, therefore, 440 cycles per second. The unit of frequency is the hertz (Hz), after Heinrich Hertz, a nineteenth-century German physicist who was the first to transmit radio waves into the atmosphere. The lowest note on a standard piano vibrates at 27.5 Hz; the highest note at 4,186 Hz. The obvious correlation is that as the frequency rises, so does the pitch of the sound. In fact, the terms *pitch* and *frequency* are often used interchangeably. They are *not* the same, although the distinction is subtle. *Frequency* is an objective quantity. You can measure it on a frequency counter. *Pitch*, however, cannot be measured in this fashion. It is subjective and can vary from person to person. Different sets of ears can hear the same sound differently. A deaf person can ascertain frequency (using a counter) but cannot discern pitch. The unit of pitch is the *mel*. I have never had occasion to use the mel: I include it here in the interest of completeness—and because it's the answer to a *Trivial Pursuit* science question.

The frequency range of the human ear is pretty wide, running from a low of 16 Hz to a high of around 16,000 Hz or, for the sake of brevity, 16 kilohertz (kHz; kilo = 1000). Ears differ greatly. I've had students who've been able to hear down to 13 Hz, and up to 20 kHz. However, the 16–16k range seems to be a pretty good average. It's also easy to remember.

Don't be overly impressed with the human hearing range. As good as it is, it represents only a fraction of the frequency range employed by science and industry every day. Radio signals on the AM band are transmitted on carrier waves vibrating at about 550 to 1,600 kHz. FM stations broadcast from 88 to 108 megahertz (MHz; mega = million). Television transmission gets in the range of gigahertz (giga = billion).

As we get older, our ability to hear higher frequencies (those over 10 kHz, for example) diminishes. This is a natural, albeit regrettable, part of the aging

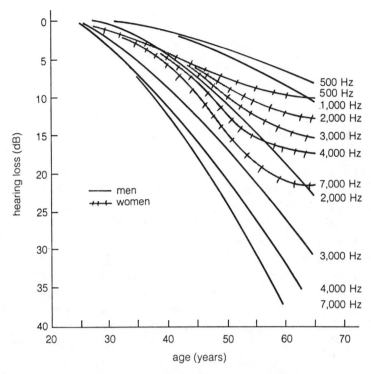

1.4 Hearing loss in men and women. Note how, although everyone starts losing some sensitivity around age 30, hearing loss occurs more dramatically and to a greater extent in men. (From AUDIO IN MEDIA Third/Second Editions, by Stanley R. Alten © 1990, 1986 by Wadsworth, Inc. Reprinted by permission of the publisher.)

process and has already begun for most people by the thirtieth birthday. These higher frequencies give music its brilliance and sharpness. Many percussion sounds, notably cymbals, sound dull and lifeless with the higher frequencies removed or decreased. For most of us, though, the process is so gradual, we barely notice the loss (see Figure 1.4). However, for those who've spent years listening to music with their amplifiers cranked up, or sitting in the front row at rock concerts, I have bad news. The high-frequency loss that begins for most of us at age 30 has begun for them probably 10 (or more) years earlier. Then when they hit 30, the natural aging process will compound the loss. Not only that, but *loudness* perception is also impaired. The music doesn't sound loud enough, so they make it louder, further damaging their ears. Eventually, this new, higher level won't be loud enough, so again the volume is cranked up, and again the damage is increased, and so on. I worked with a disc jockey whose hearing had become so impaired, he had to keep his headphones at a deafening level. It was so loud, I could hear his headphones clearly while he was wearing

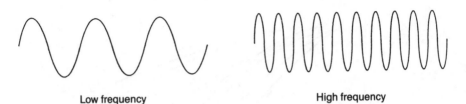

Low frequency   High frequency

1.5 Lower frequency indicates fewer cycles per interval of time. This is perceived by your ear as lower pitch.

them; and I was 10 feet away on the other side of the control room.

Before we leave the subject of frequency, one other related item should be mentioned—wavelength, the physical length (in feet, inches, millimeters, etc.) of a complete wave. Figure 1.5 shows two frequencies during 1 second of time. The lower of the two frequencies produces fewer waves per second than the higher frequency. Consequently, the length of one of the lower-frequency waves is greater than one produced by the higher frequency. Obviously, wavelength and frequency are inversely proportional; that is, as one increases, the other decreases. The higher the frequency, the shorter the wavelength. The lower the frequency, the longer the wavelength, as shown in Figure 1.6.

Wavelengths run a wide gamut. The lowest note on a piano produces a wave 40 feet long. The highest note produces a 3-inch wave. Here's a handy formula to remember:

$$S = f\lambda$$

where S is the speed of sound (approximately 1,100 feet per second), f is the frequency of any tone, and λ (the Greek letter lambda) is its wavelength (in feet). Using this formula, because the speed of sound is fairly constant, it's easy to solve for one of the variables (frequency or wavelength) if you're given the other. For example, what's the wavelength of a 1000-Hz tone?

If $S = f\lambda$, then $\lambda = S/f = 1100/1000 = 1.1$ feet

What frequency produces a 25 foot wave?

If $S = f\lambda$, then $f = S/\lambda = 1100/25 = 44$ Hz

Amplitude

Just as *frequency* and *pitch* are often used interchangeably, the same holds true for *amplitude* and *loudness*. On the sine wave of Figure 1.3, the *amplitude* is the height the curve reaches at point B (or the depth it reaches at point D). As the power of the initial vibration increases, the amount of compression and rarefaction in the wave also increases. Drop a pebble into a pond,

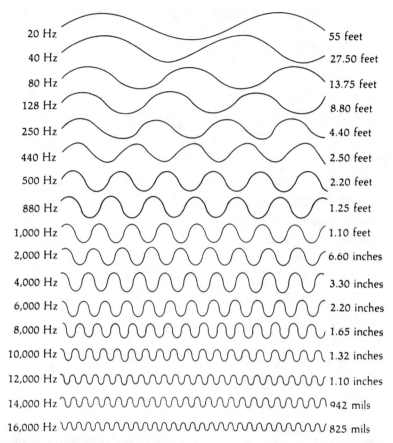

20 Hz	55 feet
40 Hz	27.50 feet
80 Hz	13.75 feet
128 Hz	8.80 feet
250 Hz	4.40 feet
440 Hz	2.50 feet
500 Hz	2.20 feet
880 Hz	1.25 feet
1,000 Hz	1.10 feet
2,000 Hz	6.60 inches
4,000 Hz	3.30 inches
6,000 Hz	2.20 inches
8,000 Hz	1.65 inches
10,000 Hz	1.32 inches
12,000 Hz	1.10 inches
14,000 Hz	942 mils
16,000 Hz	825 mils

1.6 The relationship between frequency and wavelength. As frequency increases, wavelength decreases. 20 Hz is slightly lower than the lowest note on a piano. 16,000 Hz is approximately two octaves above the piano's highest note. (From AUDIO IN MEDIA Third/Second Editions, by Stanley R. Alten © 1990, 1986 by Wadsworth, Inc. Reprinted by permission of the publisher.)

and you'll produce small waves. Drop a boulder into the pond, and the waves will be substantially higher (and the troughs deeper). The amount of force with which a sound wave strikes our eardrum is directly proportional to its amplitude: The greater the pressure of the wave against the eardrum, the louder the sound perceived by the brain. *Amplitude*, like *frequency*, is an objective, measurable quantity: *loudness*, like *pitch*, is subjective, varying from person to person. The loudness of the heavy-metal–acid–punk–new-wave music blaring from teenagers' radios may sound fine to them but may be an excruciating experience to their parents, aesthetic considerations notwithstanding. There are two units of measurement for loudness, the sone and the phon. They're not mentioned in *Trivial*

Table 1.1   A Sampling of Sounds and Their Respective Decibel Levels

Sound source	Decibels
Cannon	220
Rocket engine	195
Jet engine	160
Propeller aircraft	140
Airport runway	135
Threshold of pain; loud music	120
Pneumatic hammer; thunder	110
Heavy traffic; large truck	100
Full orchestra playing fortissimo	80
String orchestra; busy street	70
Average conversation	60
Private office	50
Subdued conversation; quiet home	40
Quiet office; recording studio	30
Whispering; leaves rustling	20
Threshold of hearing (average ear)	10
Threshold of hearing (exceptional ear)	0

Pursuit, and you'll probably never have to use them.

The unit of amplitude or sound pressure level was originally the *bel,* after Alexander Graham Bell, the man credited with the invention of the telephone. However, the bel was found to be too large a unit, making the measurement of lower amplitude sounds difficult, not unlike trying to measure a microbe with a yardstick. So after much juggling of numbers, physicists decided that one tenth of a bel (a decibel; deci = $1/10$) was an acceptable unit of measurement, and the *decibel* continues to be the unit by which we measure sound-wave amplitude. That's the good news.

The bad news is that the decibel is unlike any other measurements you're likely to encounter. Most units of measurement are immutable and unchanging. A foot is a specific length of space. A second is a specific length of time. However, with the decibel, we have a unit that represents not a single fixed quantity, but a *ratio,* a comparison of two quantities. A sound wave possesses power that is felt when the wave strikes an eardrum. Physicists calculated that the smallest amount of power that could be perceived by the average eardrum was $\frac{1}{1,000}$ of a watt, or one *milliwatt.* This tiny amount of power was dubbed the threshold of hearing and given the designation of 0 decibels (dB). Today, the decibel level of any given sound is actually the ratio of the power of that sound compared to 0 dB. (See Table 1.1 for a sampling of sounds and their respective decibel levels.)

To further cloud the issue, the decibel scale is logarithmic. Remember logarithms, the terror of trigonometry, responsible for throwing countless eleventh-graders into varying states of catatonic shock? Believe it or not, logarithms were designed to save time, especially when multiplying or dividing really large numbers. Put simply, every number, regardless of its size, has a logarithmic equivalent. When the logarithms of two numbers are added, the result is the logarithm of the product of the two numbers. For example, let's say we want to multiply two five-digit numbers. With straight multiplication, and with no calculator on hand, the process is time consuming and, if you're not particularly good at math, fraught with the possibility of error. However, with a logarithm table in hand, you quickly look up the logarithms of each of the two numbers and add the two logarithms, giving you the logarithm of the product. You check in the table to see what number has this particular logarithm—and you have the answer to the problem in a fraction of the time.

All this means is that the table of decibel levels is based on *geometric* (or multiplicative) increments, rather than *arithmetic* (or additive) increments. Each 10-dB interval on the scale represents a tenfold increase in sound level, no matter where on the scale you're looking. Ten decibels is ten times louder than 0 dB. Fifty decibels is ten times louder than 40 dB. An increase of 20 dB would result in a hundredfold increase in sound pressure on the eardrum (two increments of 10 dB). Forty decibels is 10,000 times louder than 0 dB (four increments of 10). The Richter scale that geologists use to measure earthquakes is also a logarithmic scale. Each step on the scale represents a tenfold increase in the earthquake's power. A quake that registers 7 on the scale is ten times stronger than one that registers 6, and only one tenth as strong as one that measures 8. Here's one other bit of information about the decibel scale. Because a 10-dB boost increases the amplitude of a sound by a factor of ten, to *double* a sound's level, you need to increase it by 3 dB. A 60-dB level will sound half as loud as one at 63 dB. A level of 100 dB will be twice as loud as a level at 97 dB. I know this is all rather odd, but if the human hearing range in terms of frequency is sizeable, the range in terms of sound pressure level is literally infinite. Unlike frequency, with amplitude, we have no upper limit, although if you're in the habit of regularly exposing your ears to levels above 120 dB, you can kiss your hearing goodbye. So we need a scale that can compress these enormous intervals into a form that's manageable and meaningful.

One last thing. Our decibel is referenced to a milliwatt. Technically, to further confuse the issue, there are two other decibel units, both referenced against electrical *voltages*, but, thankfully, we don't have to use them. After all, how much of this stuff can a sane human endure?

Although frequency and amplitude are the two components of sound most important to us, there are two others I want to cover briefly—timbre and envelope. These two have a much greater importance to music production than to commercial and radio production, but you still ought to know something about them.

Timbre

When you pluck the string of a guitar, if you observe the vibration of the string in slow motion, instead of a simple, back-and-forth motion, as with a pendulum, you'll see the entire length of the string jerking around in an erratic, uneven, almost violent fashion. Yet, this seemingly irregular motion produces a single, clear tone. In fact, what you hear is a *composite* tone—that is, a tone made up of dozens of individual tones, most of them too high in frequency and/or too low in amplitude to be heard by the average ear. Yet, when combined, they produce the distinctive mixture that our ear hears as the guitar's sound. Musicians refer to these tonal building blocks as *harmonics, overtones,* or *partial tones,* and it is these tones that give musical instruments their distinctive sounds.

Think about it. If you heard a note played on a piano and then heard the same pitch at the same decibel level played on a trumpet, do you think you'd have any difficulty distinguishing which tone came from which instrument? I hope not. Different types of instruments produce tones with different combinations of harmonics; and these combinations give an instrument its distinctive quality of tone, otherwise known as *timbre* (pronounced like *tamber*). A modern synthesizer electronically duplicates the sounds made by musical instruments by producing a pure, harmonic-free tone, and then allowing the operator to add any harmonic or combination of harmonics. The result is that a synthesizer can be made to sound like any musical instrument, or, using harmonic patterns that don't exist in nature, an infinite number of unusual sounds can be generated.

Envelope

Though a sound's level may appear to be uniform, in fact it may consist of a number of different levels at different points over time, particularly at its beginning and at its end. These varying levels comprise the sound's *envelope* and constitute the final way in which sounds differ. A sound's envelope has four basic parts:

1. *attack*—how long it takes to reach peak loudness
2. *decay*—how long it takes to go from peak loudness to the sustain level

3. *sustain*—the level maintained as long as the note is held
4. *release*—how long it takes for the sound to fade completely away, after the note has been released

As an example, let's compare the tones of a guitar and a clarinet.

- attack—The guitar string has a very short attack time, reaching its peak loudness almost instantly. The clarinet has a longer attack time, doing a short glide into the note.
- decay—The guitar's decay time is fairly short, but the clarinet's is almost nonexistent if the player's wind is strong and stable.
- sustain—The guitar doesn't maintain its peak loudness for long, and shortly begins fading away. The clarinet will maintain its loudness as long as the player can keep blowing.
- release—The guitar produces a long, gradual fade, as the string slowly stops vibrating. The clarinet release is much more abrupt, beginning when the player stops blowing, and ending shortly thereafter.

As you can guess, in order to perfectly duplicate the sound of a particular instrument, a synthesizer must reproduce not only the instrument's harmonic structure, but also its envelope.

As stated previously, timbre and envelope are more in the province of music production, but if you're going to fully understand the nature of sound, you're going to have to be at least marginally familiar with all its aspects.

Sound Phenomena

Now that we have a handle on what sound *is*, let's see what it *does*. Many (if not all) of these phenomena should be familiar to you (even if their names aren't):

- resonance
- Haas effect
- Doppler effect
- masking
- relative loudness

Resonance

Remember the commercial in which Ella Fitzgerald breaks a glass using only the power of her voice (and an amplifier)? The trick (and it is a trick!) depends more on the tone she sings than on her decibel level. What we're dealing with is a phenomenon called *resonance,* and it occurs when two conditions are present:

1. A sound (generally a musical tone) is produced inside a closed or nearly closed container that has parallel surfaces (a room for example).
2. A segment of the sound's wavelength (generally one fourth or one half of the wavelength, depending on the type of container) or any number of these segments equals the same length as the distance between two of the parallel surfaces.

When these two conditions occur, something strange happens. The container acts as an amplifier, and the loudness (and strength) of the sound is increased noticeably. If your bathroom has mostly tile and porcelain surfaces, try this. Go inside and close the door. Then quietly sing the lowest note you can, using an "oo" or "ah" syllable. Very slowly, as if you're playing a trombone or a slide whistle, glide up through your range, maintaining a consistent volume. Whenever you sing one of the *resonant frequencies* of the room, the loudness will jump dramatically. You'll know it when you hear it. Likewise, when you sing slightly higher or lower than one of these frequencies, the volume will drop back down. What you're hearing is the room acting as an amplifier for a *standing wave;* some dimension of the room (length, width, height) is exactly the same length as, in this case, either half of the wavelength of the tone you sang, or a whole-number multiple of this half wavelength.

Let's get back to Ella Fitzgerald and the shattering glass. If you take any glass, preferably of good quality (like crystal, for example), and you tap it gently, it will ring. This tone is the resonant frequency of the glass. If you set a speaker near the glass and use a tone generator and an amplifier to play the same tone back at a loud level, the standing wave that will be created in the glass will tear the glass apart if the amplifier is cranked up high enough. Resonance causes the glass to shatter, not a particular brand of recording tape or the marvelous voice of Ella Fitzgerald.

The need to avoid resonance is the reason that musical instruments and concert halls don't have parallel surfaces. Romantics will tell you that a guitar's hourglass shape is meant to approximate the form of a woman—cute story, but nonsense. The fact is, many of the earliest guitarlike instruments were box shaped, and therefore susceptible to resonances. Imagine serenading your lady with one of these things and having some note blast out five times louder than any of the others. The curved shape of the guitar (and many other instruments) takes care of the resonance problem.

The same is true of concert halls. The floors, ceilings, and walls are never parallel, so as to avoid these annoying, although interesting standing waves.

Recording studios are generally constructed in shapes that would drive a geometry professor crazy. The floor plan might resemble anything from a *trape-*

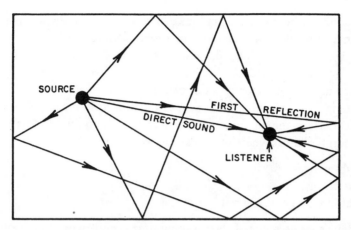

1.7 Haas effect—direct and indirect (reflected) sound. If the reflections strike the ear within 50 milliseconds of the direct sound, the reflections are perceived as being part of the direct sound. If the delay is greater than 50 milliseconds, the ear hears the reflections as echoes. (Reproduced from THE ACOUSTICAL FOUNDATIONS OF MUSIC, Second Edition, by John Backus, by permission of W. W. Norton & Company, Inc. Copyright © 1977, 1969 by W. W. Norton & Company, Inc.)

zium (a four-sided figure with no parallel sides) to a shape with six or more large sides and a few small nooks and crannies.

Haas Effect

When you're in a classroom (or any room), and you're listening (attentively, I hope) to your instructor, although it may seem that the sound you hear comes directly from the teacher, what you're actually hearing is a composite sound, made up of direct sound and indirect reflections. As the person speaks, the entire room is filled with the sound of the voice. Some of the sound comes directly from the speaker to you, but most of the sound comes to you indirectly, after bouncing off walls, floor, desks, tables, and so on (see Figure 1.7). Obviously, the direct sound reaches you much more quickly than the indirect sound, which has been bouncing around the room. With most of the sound coming from all directions, why do we only perceive a single sound, coming from the speaker? A fellow named Helmut Haas figured it out in the early 1950s, although other people had been working on the problem for some time. Nevertheless Haas' name is the one forever linked with this effect. What he found was that if the indirect sound hits your ear within 40 or 50 milliseconds of the direct sound, the brain perceives the whole package as direct sound, and it will seem to come from only the single source—in this case, the teacher.

Should some of the indirect sound take longer than 50 milliseconds to reach your ear, as it would in a large room, or a room with reflective surfaces, it will be perceived as echoes. This, like resonance, is a major concern to architects who design auditoria and concert halls because, as you can imagine, echoes can ruin a musical or dramatic performance for an audience.

As long as we're talking about echoes and how to eliminate them, we need to touch briefly on the subject of sound reflection and absorption. Most materials do one of three things when hit by a sound wave:

1. absorb it
2. reflect it
3. absorb part of it and reflect the rest

Acousticians spend much of their waking time dealing with sound absorption and reflection. I don't. Still, you ought to know a few things about the subject.

All building materials (steel, stone, wood, cloth, tile— everything) have been given a number called an *absorbtion coefficient*. This number, which can range from 0 to 1, indicates the degree to which the material absorbs sound. A 0 (zero) rating means no absorption; The sound wave hits the material and bounces off, like a cue ball hitting the rail of a pool table. A coefficient of 1 indicates complete absorption—no reflection at all. The sound is sucked up like water in a sponge. Metal, ceramic tile, and polished stone all have low coefficients. Cloth, foam rubber, and people have high coefficients. (See Table 1.2 for an array of materials and their respective absorption coefficients.)

I can't think of anything that scores a perfect 0 or 1. Everything falls somewhere in between. In addition, most substances have different coefficients for different frequencies or frequency ranges. Frankly, I'd just as soon leave this whole thing to the acousticians, although to be honest, the acoustical makeup of your recording studio is of paramount importance. In most studios, the philosophy is "the deader the better." You don't need echoes bouncing around when you're trying to record. Most producers I know record their tracks dry—no echoes or reverberation (reverb)—and if they so choose, they'll add echo or reverb later.

You should know that echo and reverb are not the same thing. *Echo* is a discrete, distinct repetition of a sound—like when you stand on a mountain and yell "hello" and a clear "hello" comes back at you. *Reverb*, on the other hand, is multiple echoes. Reverb is what gives a room its character and conveys to the ear the room's size. Stand in a large, empty room with your eyes closed, clap your hands once, and listen. The sound bounces around the room like a slew of berserk billiard balls, until, after a very few seconds, it fades away. This collection of echoes is reverb, and it's what makes or breaks concert halls. Many performers, when they check out the acoustics of a hall, do the same clap test, and listen to see how the sound decays (fades). If the decay is too long, echoes may be a problem, if the decay is too short, the room may sound

Table 1.2   Absorbtion Coefficients

Material	125	250	500	1,000	2,000	4,000
			Frequency (Hz)			
Acoustical plaster, 1 inch	.25	.45	.78	.92	.89	.87
Asphalt tile on concrete	.02	.03	.03	.03	.03	.02
Audience in upholstered seats	.60	.74	.88	.96	.93	.85
Brick, unglazed	.03	.03	.03	.04	.05	.07
Brick, unglazed, painted	.01	.01	.02	.02	.02	.03
Carpet, heavy, on concrete	.02	.06	.14	.37	.60	.65
Carpet, heavy, on felt	.08	.27	.39	.34	.48	.63
Chairs, metal or wood, occupied	.15	.19	.22	.39	.38	.30
Conrete, unpainted	.01	.01	.01	.02	.02	.03
Floor, wood	.15	.11	.10	.07	.06	.07
Glass, heavy plate	.18	.06	.04	.03	.02	.02
Glass, window	.35	.25	.18	.12	.07	.04
Marble or glazed tile	.01	.01	.01	.01	.02	.02
Plaster on lath on studs	.30	.15	.10	.05	.04	.05
Plywood on studs, ¼ inch	.60	.30	.10	.09	.09	.09
Velour drape, heavy	.14	.35	.55	.70	.72	.65
Velour drape, light	.03	.04	.11	.17	.24	.35

lifeless and unnatural—more for the acousticians to worry about. Echo and reverb are important aspects of production—more about them later.

Doppler Effect

Everyone has experienced the Doppler effect (named after Christian Doppler, a nineteenth-century German scientist), but most don't know how to explain it. Remember the "Road Runner" cartoons? Amazingly, there was only a single story line—for all of them. Wile E. Coyote would try to capture the Road Runner, using schemes and devices that ranged from the ludicrous to the positively ingenious—only to find himself repeatedly falling off a precipice, and plummeting to the canyon floor far below. The sound made by the coyote in his descent is a perfect example of the Doppler effect (and your parents thought "Road Runner" cartoons were useless—Hah!). Imagine you're standing on the cliff as the coyote falls off. What happens to the pitch of his scream as he falls? It drops. If we could somehow ride next to the coyote in an elevator as he de-

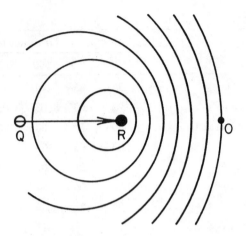

1.8 Doppler effect—As R moves rapidly toward O, the sound waves between R and O become compressed, and O perceives a *rise* in pitch. At the same time, the waves between R and Q become further apart, causing Q to hear a *drop* in pitch. (Reproduced from THE ACOUSTICAL FOUNDATIONS OF MUSIC, Second Edition, by John Backus, by permission of W. W. Norton & Company, Inc. Copyright © 1977, 1969 by W. W. Norton & Company, Inc.)

scends, would the pitch of his scream change? Not at all! What's happening here, as Doppler pointed out, is that as the coyote falls and his speed increases dramatically, the sound waves between him and the ground become compressed, and the waves behind him—between him and the top of the mountain—are stretched apart. Remember the correlation between wavelength and frequency? They're inversely proportional: as one goes up, the other goes down. As the waves behind the coyote are stretched further apart, the wavelength obviously increases. As the wavelength gradually increases, the pitch becomes progressively lower. That's what we hear. If we were to stand on the ground directly under the falling coyote (well all right, a little off to the side), because the waves between him and us are being compressed, the wavelength is getting shorter. Therefore, standing on the ground, we'd hear the pitch gradually rise. Think of a race car going by you. As it approaches you, the waves in front of it are compressed, so the pitch rises. As it zips by you, the waves behind it are elongated, so the pitch drops. (See Figure 1.8 for an illustration of the Doppler effect.) Thank you, Herr Doppler.

Masking

Here's an interesting phenomenon. You're in your car on a warm summer day, doing 30 mph as you head for the interstate highway. The window is open, and you've turned your radio up, so that if it were any louder, it would be uncomfortable. When you reach the highway and accelerate to 55, what happens? With the added wind and engine noise, the radio is hard to hear, so you turn it up—to a level that should be painful, but isn't. When you get off the highway, the first thing you do is turn down the radio, which is now blasting away, although moments ago it was just fine.

What you've experienced is a phenomenon called *masking*. It occurs when you combine two sounds of differing frequencies. Theoretically, if one of the sounds is very loud, the addition of the second should make the overall level considerably louder. So much for theory. In this case, the lower frequency masks or covers the higher frequency: the low frequency wind and engine noise will mask the higher frequency of the radio. This effect is analogous to seeing car headlights in your rearview mirror. At night, the lights from the car behind you can be blinding; in the day, with the masking effect of the sun, you'd hardly notice them.

Relative Loudness

Your ears are not democratic: They do not hear all frequencies equally. Listen to two tones (25 Hz and 1 kHz, for example) recorded at the same decibel level. Although a meter may show the levels to be identical, your ear will hear the higher frequency as considerably louder. Two fellows named Fletcher and Munson were the first to really document this situation, constructing an odd set of graphs that we still use today, called, appropriately enough, the *Fletcher–Munson curves* (see Figure 1.9). They show that as the frequency rises (at a constant decibel level), the ear will perceive the sound as becoming louder—up to around 3 kHz. Above this, the sound appears to diminish in volume. This sensitivity to higher frequencies, particularly in the 2- to 4-kHz range has prompted a rash of dubious claims, purporting to place a number of obnoxious sounds (a baby's cry, a woman's scream, the sound of fingernails on a blackboard) within this most sensitive section of our hearing range, thus, supposedly, accounting for our aversion to them. I think these notions are silly, but they do make for good conversation when you need a topic.

Summary

Whatever you do, don't assume that this chapter resembles anything remotely approaching a complete analysis of sound. I only wanted to accomplish two things:

1. To give you an overview of what sound is and does—You can't create unless you know your medium.

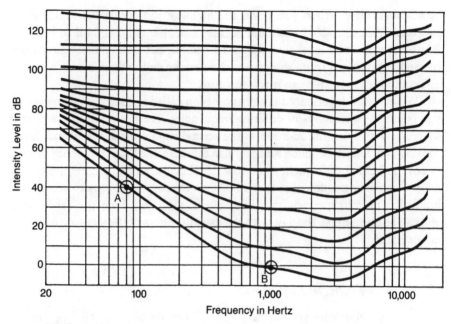

1.9 Relative loudness curves (based on Fletcher–Munson). An 80-Hz tone at 40 dB
(A) sounds as loud as a 1-kHz tone at 0 dB (B). (From AUDIO IN MEDIA Third/
Second Editions, by Stanley R. Alten © 1990, 1986 by Wadsworth, Inc. Reprinted
by permission of the publisher.)

2. To prove that there's nothing terrifying about fundamental physics—
   In fact, all I've done is to explain aspects of sound with which I'm sure
   you're already familiar. Anybody can grasp this—just as *anybody* can do
   creative radio production.

Review

1. Sound consists of a vibration passing through an elastic medium to a
   receiver. As the medium's elasticity increases, so does the speed of the
   sound. As the medium's density increases, the sound's speed decreases.
   The speed of sound is 1,087 feet per second at sea level, at 32 degrees
   Fahrenheit.
2. The human ear consists of three chambers: the outer, middle, and inner
   ears. The auditory nerve passes vibration to the brain, where the vibration is perceived as sound.
3. Sound waves can be either longitudinal or transverse.

4. Waves consist of alternating areas of compression (high pressure) and rarefaction (low pressure).

5. *Frequency* (number of cycles per unit of time) is measured in hertz (Hz). The human hearing range is around 16 Hz to 16 kHz. Frequency and pitch are directly proportional—an increase in one corresponds to an increase in the other.

6. *Wavelength* is the distance between the identical points of adjacent waves. Frequency and wavelength are inversely proportional—an increase in one corresponds to a decrease in the other.

7. *Amplitude* is a measure of sonic power and is measured in decibels. On a logarithmic scale, 0 dB is the threshold of hearing, and 120 dB is the threshold of pain.

8. *Harmonics* are partial tones, which, when added together, provide a musical instrument's *timbre* (tone color).

9. A sound's *envelope* consists of its attack, sustain, and decay times.

10. Resonance is a phenomenon in which a specific portion of a wave fits its container precisely and becomes amplified by the container.

11. The Haas effect occurs when direct sound and its reflection strike the ear within 50 milliseconds (msec) of each other, and the ear perceives them both together as direct sound. If the delay is greater than 50 msec, then the ear hears an echo.

12. Materials can reflect or absorb sound. The degree to which a substance absorbs sound is shown by its absorption coefficient.

13. The Doppler effect occurs when a sound source traveling at high speed stretches or compresses the wavelength of its sound in relation to the receiver, resulting in a perceived pitch change.

14. Low-frequency sounds can mask high-frequency sounds.

15. The human ear doesn't hear all frequencies within its range equally. The most sensitive portion of the human hearing range is between 2 and 4 kHz.

2

□ □ □
□ □ □
□ □ □

The Console

I offer three principles to people learning to operate production equipment.

Principle 1: It's not imperative for a good producer to be an electronics wizard.

Granted, it's a plus to really know the insides of your equipment. Let's face it, you can be an excellent driver without knowing how to overhaul your transmission. It's the same thing here. Knowing *what your gear does* is considerably more important than knowing *how it works*. You need not wade through a quagmire of numbers, graphs, spec sheets, and formulas to gain facility with your tools, any more than a carpenter needs to be proficient in metallurgy to productively use a hammer.

Principle 2: The goal of production is not to operate machines, but rather to use those machines as a means to some creative end, be it a commercial, a station promo, a play, a song, or any other on- or off-air feature.

This idea is very important. Pushing buttons doesn't constitute production. A chimpanzee can push buttons. However, a chimpanzee can't translate an intangible thought into an electronically generated sound representation. That's your ultimate goal in production; to take sounds from your brain and put them onto tape so that other people can hear them.

Principle 3: Despite the intimidating appearance of a lot of audio equipment, these devices are (for the most part) designed to operate in a perfectly logical fashion.

In other words, most of the machinery you're likely to run across, despite its apparent complexity, is really quite simple to use. Imagine a language that has only a few simple rules and no exceptions to those rules—ever. That's what you have with most studio apparatus.

Standard Functions of the Console

The production console (or *board*, as it's referred to in the trade) is to the audio producer what the heart and brain are to the human body. Despite the wide variety of consoles on the market, all of them, from the simplest to the most complex, basically perform only a few standard functions:

1. *amplification*—raising and lowering a signal's level (the *signal* is an electrical representation of a sound)
2. *mixing*—combining signals from different sources into a single compound signal
3. *switching*—sending a signal from one piece of apparatus to another

Some consoles, especially larger ones, are equipped for one other function:

4. *processing*—electronically altering the signal

That's it! Even the room-size consoles in music studios and film-production studios primarily perform only these few operations. As a result, if you *really* understand the operation of a simple small console (Figure 2.1), you'll find most of the same principles will work with a big multichannel board.

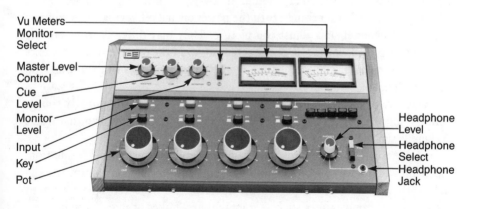

2.1 Small console. (Courtesy of Broadcast Electronics, Inc., Quincy, IL.)

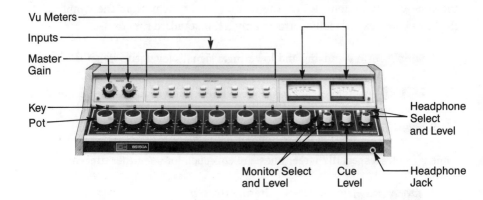

Vu Meters
Inputs
Master Gain
Key
Pot
Headphone Select and Level
Monitor Select and Level
Cue Level
Headphone Jack

2.2 Medium console. Note how the basic components are the same as on the smaller board. (Courtesy of Broadcast Electronics, Inc., Quincy, IL.)

The key word here is *principles*. Of course, the more complex the console, the more tricks it's capable of performing, but the operating principles from one unit to another are remarkably similar. In a way, it's like television sets. Most TVs allow you to do four things: turn the power on and off, regulate the volume, change the channel, and adjust the picture. They may differ as to where the controls are placed, or the size of the screen, or even the number of adjustments that can be made, but they're basically all the same. If you can run one, you can run them all.

One key to successful console operation hinges on an astoundingly simple concept: *follow the signal's path*. Over my console I have a large sign that reads: "Where is the sound coming from? Where do you want it to go?"

More than 90% of the problems fledgling producers encounter (aside from mechanical glitches) can be solved by keeping in mind this notion of the signal's path. Remember that the console operates two ways: (1) signals are sent *into* it (input) from sources such as microphones (mikes or mics), tape devices, and turntables; and (2) they're sent *out* of it (output) to speakers, to tape units, and, if you're broadcasting, to the transmitter. Let's examine the input and output chains of a typical, broadcast/production console (Figure 2.2).

The Input Chain

A signal enters the console through an *input*. Consoles have two types of inputs, high- (or line-) level and low- (or mike-) level. Signals originating from microphones or turntables are extremely low level—so low, in fact, that if you sent one through the console's amplifier and turned it up as high as possible, you probably couldn't hear the signal. So before this signal reaches the main amplifier (amp), it passes through a *preamplifier* (or *preamp*), which boosts the signal enough so that after it then passes through the main amplifier, it's loud enough to be heard and used. Tape devices (reel-to-reel, cassette, and cartridge) produce much greater signals, and need no preamplification. They enter the console through *high-level* inputs. The various inputs (high- and low- level) are connected to *input selectors*, which are usually buttons or switches positioned near the top of the console (see Figure 2.2). Each of these selectors is labeled according to which piece of equipment is wired to it. From the input selector, the signal is routed to one of the board's *channels* (sometimes called *mixers*). The channel generally consists of a rotary or slide *pot* (short for *potentiometer*, also called a *fader*), and a key switch (usually simply referred to as a *key*), often positioned just above the pot. The key performs two functions:

1. It sends the signal to one of the board's output lines (sometimes called *bus lines*).
2. It opens the channel so that the signal can reach the pot, which raises or lowers the signal's level.

Most boards have two or more output lines, so that you can have two or more signals going through the board simultaneously without their bumping into each other. For example, let's say you're playing a record, and taping it onto a cassette. At the same time, using a different internal line, you can record your voice onto a tape recorder and not worry that the two signals (music from the record and voice from the mike) will mix. These output lines are generally given names like *program line* and *audition line*. If you have three lines available, the third might be called *utility* or *auxiliary*. These lines are identical to each other, so their names have no real meaning. They merely serve to distinguish one line from another. The key assigns the signal coming from the *input selector* to a particular line, and then the signal flows to the pot. *Note*: In place of a key switch, some boards have a separate *line selector*, which only chooses the line circuit, and a separate on/off switch opens the pot.

Following the path, we see that the signal travels from a source:

1. to an input
2. to a key (or a line selector and on/off switch)
3. to a pot

This is the *input chain,* the successive console features that *must* be engaged in order to get an input signal out of the board. If any one of the three isn't on or isn't properly engaged, the signal won't get through the board.

Consoles—even small consoles—can differ greatly from one another: Some have rotary pots, and others slide pots; some have the input selectors positioned directly over the appropriate channels, and others don't; the key on some is a switch, on others a button. On all of them, however, the basic input chain is the same.

On some boards, one input selector (and consequently one piece of equipment) is assigned to each channel. However, if you have more gear than there are channels on the board, some channels obviously are going to have to accept more than one input. For that reason, most boards come with at least two input selectors wired to each channel.

The Output Chain

Once a signal has been brought into the board and sent to a pot, it begins the journey out of the board. On your board, there should be a separate control labeled *MASTER OUTPUT* or *MASTER GAIN* (refer to Figure 2.2). Unlike the pots, which control the outputs of individual channels, the master controls the output of the entire board. This is obviously a handy tool to have if, for example, you have to fade three or more pots simultaneously, especially rotary pots. With slide pots, depending on the size of your hands and the distance between the pots, you can often control two or three with each hand. Even with sliders, however, your fade will probably be smoother if you use the master.

Vu Meter

Along with the master, the most important link in the output chain is the *vu (volume units) meter.* It tells you at a glance the level of the signal leaving the master. More bad production can be attributed to misreading or not reading the meters than to any other cause. The left and right meters in Figure 2.3 are the industry standard. On each meter, you can see two scales, one for modulation percentage (ranging from 0 to 100%) and the other for volume units (ranging from –20 to +3 Vu). *Volume units* are subjective loudness units (like sones and phons from Chapter 1). The modulation percentage indicates the percentage of the channel's capacity being used. The meter is designed to react to changes in sound level as your ear would, although not nearly as quickly.

The red portion of the scale (above 0 Vu) lets you know that your signal is too loud, and in danger of distorting. However, if, in your endeavor to stay out of the red, you keep your signal too low, your signal-to-noise ratio will drop. Electronic circuits have noise in them—not a lot, but enough to be annoying

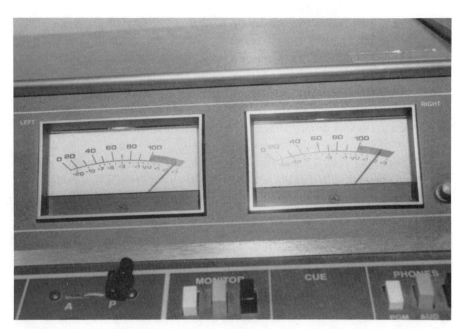

a

b

2.3 (a) Vu meter is too hot ("in the red"); (b) Vu meter is too cold ("in the mud").

if they're not masked. Turn an amplifier on, and crank it all the way up. The hissing you hear is the noise I'm referring to. The amount of noise in a circuit is constant if the amount of current flowing through the circuit doesn't change. Obviously the louder the signal you're using, the less likely you are to hear the noise. The higher the *signal-to-noise ratio* (signal level divided by noise level), the better. As the level of the signal drops, and the noise level remains the same, the ratio will drop, resulting in a poorer, noisier signal. So keeping the Vu needle too far out of the red just creates another problem.

Novices correctly assume that a good place to keep the Vu needle is right on the demarcation line at 0 Vu. However, they soon see how impossible this level is to maintain. A human voice, for example even when speaking in a natural fashion, has a dynamic range, producing sounds of various decibel levels. To keep the meter at 0 Vu, or any preset level, you'd have to talk without changing your loudness, and that's virtually impossible, not to mention unbearably boring.

Although you can't maintain a single Vu level for long, you can, however, do the next best thing—keep the bulk of your recording within an acceptable *range*. You should try to keep the needle within the 80–100% modulation area (between –2 and 0 on the Vu scale). If the needle spends most of its time within this range, you'll have a good overall level. If the needle *occasionally* pops into the red, *don't worry!* Good recording tape is manufactured to handle transient bursts of sound like this without distortion. Just don't overdo it. If the needle seems to hover consistently around +2 Vu or higher, you can bet you're distorting the signal.

The number of meters on your board depends on two factors:

1. the number of output lines
2. whether the board is wired for monophonic ("mono") or stereophonic ("stereo") sound

In a mono board, there will probably be one meter for each line. In a stereo board, there can be either two meters (one for the left side and one for the right side, as shown in Figure 2.3), or a set of two meters for each output line. So if the number of meters on a board equals the number of lines, it's probably a mono board. If there are twice as many meters as lines, it's a stereo board. Some consoles have meters for each line and a separate meter for the total output of the board—sort of a master Vu, and some boards have the capability of switching to either mono or stereo.

Meters are also critical in troubleshooting. Because all tape machines and consoles use meters, and because meters reflect the presence of a signal, you can use your meters to follow the path of a signal from device to device. For example, let's say you're playing a cassette, and you want to bring it up (send the signal) through the board. You roll the cassette, and you hear nothing. Where's the problem? Check your meters as you follow the signal's path:

SOURCE to INPUT CHAIN to OUTPUT CHAIN to FINAL DESTINATION

In this case:

CASSETTE DECK to BOARD INPUT to BOARD OUTPUT to SPEAKER

Are the meters on the cassette deck moving? If not, then the signal isn't getting out of the deck; something's wrong there. If the deck meters are moving, but the console meters aren't, then the problem is in the board, possibly in the input chain (input select, key, and pot). If all the meters (deck and board) are moving properly, but you still don't hear anything, the problem is somewhere within the output chain. Your board will have a headphone jack to monitor outgoing signals. Plug a set of phones in and listen. If the signal is coming through the headphones, but not out the studio loudspeakers, then the problem has to do with the speakers.

Remember: Follow the meters in the order in which they occur along the signal's path.

Before we leave the subject of Vu meters, let's talk about *setting your levels.* Before you begin *any* recording, you need to align all the meters in your studio. This will ensure uniform levels entering and leaving not only the console, but also all other pieces of apparatus in your system. If a signal leaves the console at 0 Vu to be recorded on a tape, you want the tape recorder meter also to read 0 Vu. Even if the console level is fine, if the recorder level is too high or too low, the level of the signal being transferred to tape will similarly be either too high or too low.

To set the levels properly, do the following:

1. Set the master output pot to a convenient position (halfway up, for example).
2. Play a tone through the board, and raise its level until the meter reads 0 Vu.
3. Set all other pieces of gear into record mode, and adjust their meters so that the tone coming from the board makes all these meters read 0 Vu.

What you've done is align all the equipment meters to the console meter. Whatever the console Vu meter indicates will also be indicated on all other meters.

Monitor System

The final link in the output chain is the *monitor system,* the system through which you can hear (monitor) what's coming out of the board. It has three elements:

1. the studio monitors (loudspeakers)
2. the headphone jack
3. the cue circuit

The monitors and the headphone jack are wired to the master output, so whatever comes out of the board also comes out of the speakers and headphones. The only exception arises when a mike channel is open. When an open mike and its speaker are too close together, the signal from the mike comes out the speaker, is picked back up by the mike, refed to the speaker, refed out the speaker, and so on, producing *feedback*, the high-pitched squeal you may fondly recall from high school assembly programs. To guard against this, most consoles have a **muting circuit** built into their mike channels. This circuit is generally tied to the channel's key switch (or the channel's on/off switch if the board doesn't use keys); so when the channel is opened for mike use, the speakers are automatically muted. With no signal coming from the speakers to be recycled into the mike, there's no feedback. The muting has no effect on the headphones (*phones*), however, since the signal coming from the phones is generally too low level and too far from the mike to be a problem—unless of course you're in the habit of cranking your phones up when you use them near an open mike, in which case, "Welcome to Let's Make 'em Squeal!"

Your board should have a *monitor volume* or *monitor gain* control to regulate the output of the speakers and a *headphone volume* or *headphone gain* control to regulate the output of your headphones.

> **Important point:** These monitor controls affect *only* the loudness coming from the speakers and headphones: they have no effect on the level coming from the board.

Somewhere near your monitor regulators should be the *monitor select* controls. Usually buttons or switches, these determine what internal line of the board (program, audition, utility, auxiliary, etc.) is fed to the monitor system.

The *cue circuit* is the third element of the monitor system. It's designed to allow you to monitor a signal while keeping the signal inside the console and away from the speakers. Each channel pot on your console has a *cue position*: with a rotary pot, the cue position is reached by turning the pot completely counterclockwise. With a slide pot, drop it all the way to the bottom. (Most pots, rotary or slide, will click when they hit cue position.) The output of that particular channel is then sent to a small amplifier and speaker inside the board. The cue system is so named because, for example, it allows you to cue up a record (that is, place the needle at the start of the song) while another disc or tape is playing over the air.

You should have a *cue monitor* control somewhere on the board. This sets the cue level coming from the little internal speaker. You need a cue monitor control because the pot, when fixed in cue position, isn't capable of changing the signal level being sent to the cue amp. Also keep in mind that the cue monitor control only affects the signal coming from the cue amp. It has no effect on the signal coming into the board. You may also have some kind of *cue channel select* switch. This allows you to send different output lines (Program, Audition, etc.) into the cue circuit.

Somebody (whose name I've long since forgotten) once told me that production is the art of finding yourself boxed into a corner and then figuring a way out—and there's almost always a way out. Try this.

Problem: You're on the air during a major blizzard when the superintendent of schools calls in with a list of school closings. You record the message onto a reel of tape, intending to air it when the record you're playing is over. You put the console pot for the recorder into cue mode and listen as the tape rewinds. Suddenly the cue amp goes dead, and you have no idea where on the tape the superintendent began talking. How can you find the start of the message?

Solution 1: If you're broadcasting on your console's program line, send the tape signal over the audition line. That'll keep the signal off the air. You'll also have to change the monitor select control to audition so that you can hear what's coming over that line. Cue up the tape, switch the tape channel output and the monitor select back to program, and you're in business.

Solution 2: This is not as exact a solution, but it's better than nothing. Watch the meters on the tape deck. If they're moving, then the tape is playing the message (inside the deck). Slowly rewind the tape and stop when the meters stop moving. You should be at or near the start *head* of your recording.

Miscellaneous Controls

Although small and midsize production and broadcast consoles contain all or most of the preceding features, it only makes sense that manufacturers would add special gadgets to make their products distinctive. Here are a few items you may encounter.

Intercom or **Talkback**—allows you to speak to and hear someone in another room, using a small internal mike in the board, and a small speaker. Sometimes, the cue speaker doubles as an intercom speaker. Most recording studios have the performance area separated from the recording area by panes of glass, and communication between people on both sides of the glass is essential. In some cases, hand gestures will suffice, but it always helps to be able to speak and be heard directly.

Phone feed—allows you to send the output of the board into the telephone. You may also have the phone line tied to a console input, so you can bring the phone signal into the board and then record it or put it on the air. Using the phone feed, you can play a client's commercial for the client directly from the control room, to check that the spot is okay. Being able to record off the phone is important if you do interviews (just ask your news department). Also, if you want to recreate a telephone conversation for a commercial or promo, it's a lot easier recording a voice directly off the phone than running a clean voice through a processor to simulate a telephone sound.

Tone generator—a device that generates a tone (usually 1 kHz) within the board. It provides a reference level against which you can set the vu levels of your other pieces of equipment.

Variable input—options for input channels. Whereas most inputs are wired directly to a single channel (key and pot), a variable input gives you a choice of two (sometimes more) channels to which a signal can be sent.

Network input—for on-air boards. The feed from a network is tied directly to this specific input. In the absence of this feature, the network line can be tied to any existing console input.

Remote starts—a must for large studios. Allows you to start other pieces of apparatus (recorders, turntables, etc.) from the console.

Timer—important when you're producing material that has time constraints, such as commercials, which have to stay within 30- and 60-second limits. The timer can be triggered directly, or it may be tied into the start-up circuit of another piece of equipment. For example, if most of your commercials are recorded on cartridges (carts) (see Chapter 4), the timer might be tied to the cart machine. So when you're ready to record something, the triggering of the cart machine also resets and starts the timer.

These are *not* standard features, but many of them, if not already on the board, can be built in by your engineer. Most consoles (especially the really expensive ones) have their own distinctive doodads. Some are useful, some merely fluff designed to boost the price. Nonetheless, all small and midsize boards utilize input chains and output chains that are set up exactly like, or at least similar to, the systems discussed here. Remember: if you can run one, you can run them all.

Large Multichannel Recording Consoles

If you've ever seen one of those huge professional multichannel boards, it probably looked like a dizzying array of meters and switches and knobs. Don't panic. Big boards do what small boards do, but on a larger scale, and with more options.

Closer examination of a large board reveals a number of things.

1. There are two distinct sections of the board: (1) the input section on the left, possibly extending to (or past) the center; and (2) the output section on the right.
2. The controls are arranged in vertical modules. *Input modules* (see Figure 2.4) control and shape signals coming into the board from outside sources. *Output modules* send the signals to the monitor system and to specific tracks on the recording tape.
3. The input modules (Figure 2.4) appear identical to each other.
4. The output modules appear identical to each other. Some large consoles have separate monitor modules.

Like small and midsize consoles, big boards are very similar to each other in features and controls. Obviously, the bigger the board, the more gadgetry you can expect. Here's an overview of some of the input and output devices you're likely to encounter. They're arranged as they would most probably occur, starting at the top of the module, but expect variation from manufacturer to manufacturer and from model to model.

Input Controls

Mic/line select—allows selective treatment of high- or low-level incoming signals. Since a mike signal is low level, this control, when set to mic adds needed preamplification. When the incoming signal is from a high-level source, like a tape recorder, no preamplification is needed, so the line setting accepts the signal and sends it directly to the main amplifier.

Pad—automatically lowers the level of an incoming signal a preset amount. A pad protects against distortion or overload when the signal is too *hot* (loud). A pad has only two settings—on or off—and, depending on the board, it will cut the volume anywhere from 10 to 50 dB.

Trim—similar to a pad, however, the trim control is variable, allowing you to gradually attenuate (lower) the level.

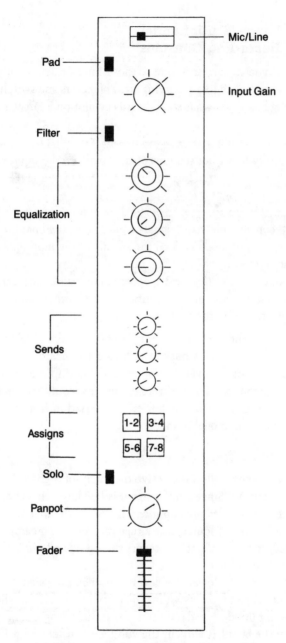

2.4 Typical input module from a multichannel board.

Input gain—functions like the trim control. You can also raise, as well as lower, the level of the incoming signal.

Filter—cuts off a portion of the signal's frequency range, usually on either the upper or the lower end.

Equalizer—breaks the signal's frequency range into high-, middle-, and low-frequency bands. It allows you to raise or lower the level of certain frequencies within these bands. Equalization is the key to creating a sound's color and texture (especially in music).

Auxiliary sends—route the signal to a variety of destinations: the cue circuit, the monitor circuit, special effects and processing devices, etc.

Bus assigns—route the signal to any of the output modules.

Panpot (short for *panoramic potentiometer*)—places a stereo image at any point between the speakers (left side, right side, center, left of center, right of center, etc.) Though we won't deal with stereo until chapter 12, you should be aware that the space between your speakers is like an artist's canvas, and you can put sound images anywhere within this field.

Solo—allows you to isolate and hear one channel, by muting all other channels *within the monitor system only*. Note that the output of all the channels to the output module(s) isn't affected.

Fader—controls the signal level sent from a channel to an output module.

Output Controls

An output module sends the signals from the input modules out to the monitor system and to tape recorders. Consequently, the output module is a much simpler device than an input module, with far fewer features. Here are some of them.

Input control—varies the combined levels coming from the input modules.

Return controls—allows some output modules to send signals from the tape recorder through the monitor system. In this case, there would be some sort of input selector, letting you choose which source (input modules or tape recorder) you want to send to the monitor system.

Solo—This is the same as on the input module.

Panpot—This is the same as on the input module.

Fader—This is the same as on the input module.

Monitor Controls

If you have a separate *monitor module*, it'll probably contain these features:

Monitor fader—fades the monitor system without affecting the actual output of the board.

Talkback system—like an intercom. It allows you to communicate with performers in the studio while you're in the control room.

Cue send—sends a signal into the cue speaker, just like the cue control in the smaller boards.

Headphone jack and **Volume control**—allows specific volume control for the headphones that have been plugged into the jack.

Monitor pan—adjusts the monitor output in terms of stereo placement (left, right, center, etc.).

Although some of these consoles appear intimidating, it shouldn't take you long to acclimate to any of them. Most radio stations don't need the capability of a huge board to produce their commercials, station promos, and on-air features. This, coupled with the hefty price tag of a large console, means that small and midsized consoles are the order of the day in all but the largest and wealthiest stations.

Review

1. Knowing how to operate equipment properly is more important to creative production than knowing how the equipment works. The gear is merely a means to an end.
2. Most equipment is designed to operate simply and logically.
3. A console, regardless of size, performs only three main functions: amplification, mixing, and switching.
4. Always be aware of the signal's path. Where is the sound coming from? Where do you want it to go?
5. A signal moves through the console via the *input chain*—console input (high or low level), key switch, and pot.
6. The signal leaves the console through the *output chain*—bus line, master gain, vu meters, and monitor system.
7. The *monitor system* consists of the studio monitors, the headphone jack, and the cue circuit.
8. The vu meter is central not only to maintaining proper level, but also as a troubleshooting tool.
9. Align all meters before recording (i.e., set your levels).
10. Large multitrack consoles consist of input and output modules. The input modules have identical controls, as do all the output modules.

3

□ □ □
□ □ □
□ □ □

Microphones

Compared to consoles, the rest of the equipment we cover in this book is considerably less complicated and therefore much easier to learn about. A microphone (often shortened to mic or mike) is a fascinating tool. It's a *transducer*, a device that changes energy from one form to another. In this case, *acoustical energy* (sound) is changed into *electrical energy* (the current that flows from the mike). Here's how it works: the mike is equipped with a diaphragm, a small, thin, sensitive membrane. Because it's so sensitive, sound, even faint sound, can cause the diaphragm to vibrate. As the diaphragm vibrates, an electrical current is produced, varying with the movement of the diaphragm. The louder the sound, the greater the pressure on the diaphragm. The greater the pressure on the diaphragm, the greater the electrical current generated by the mike. As a result, the current emerging from the mike is actually a series of electrical pulses corresponding to the movement of the diaphragm; since the movement of the diaphragm corresponds to the pressure of the sound, it stands to reason that the current also corresponds to the sound's pressure. It's like translating a poem from one language to a second, and from the second to a third. Theoretically, linguistic differences notwithstanding, the poem in the third language should parallel the original.

If this seems a little fuzzy, consider a telegrapher. He or she takes a word, changes it into a series of dots and dashes, and then sends these dots and dashes via an electrical current through a wire. The person on the other end receives the dots and dashes and then decodes them back into the original word. Sound (acoustical energy) causes the diaphragm to vibrate (change to mechanical energy), which causes the current to flow (change to electrical energy), which causes the speaker to vibrate (change back to mechanical energy), which re-creates the sound (change back to acoustical energy). (See Figure 3.1.)

Aside from physical appearance, mikes differ in two respects:

1. how they generate current
2. their response patterns—that is, the direction(s) from which the mike is most sensitive to incoming sound

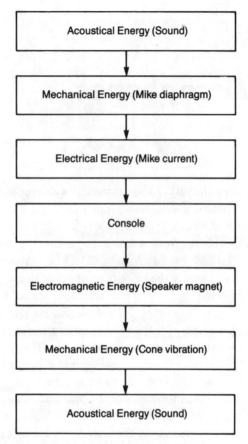

3.1 The various energy changes (transductions) a sound undergoes from when it first hits a mike to when it emerges from a speaker.

Types of Microphones

Moving-coil Microphones

Perhaps the most common type of microphone, the moving coil mike (sometimes referred to as a *dynamic mike*; see Figure 3.2); generates electricity using a principle you may remember from school (or from Mr. Wizard). When a coil of wire vibrates within a magnetic field, electricity is *induced* (made to flow) through the wire. In the moving-coil mike, the diaphragm is connected to the coil of wire (sometimes called a *voice coil*), and the coil surrounds a magnet. When the diaphragm vibrates in reaction to sound, the coil moves in a similar fashion around the magnet, causing a small current to flow through the coil and out the mike.

There are pluses and minuses to every mike and every type of mike. With the moving-coil mike, the pluses greatly outweigh the minuses. The moving-coil mike is relatively inexpensive, very durable, and responds well to powerful signals, such as those created by screaming singers. Unlike condenser mikes, the moving-coil mike doesn't require a separate power source, and so it's quite portable. The only thing less than terrific about this mike is its frequency response, especially in the high-frequency range. (A mike's *frequency response* is the range of frequencies to which it's capable of responding.) For example, a decent vocal mike might have a range of 100 Hz to 10 kHz. Actually, it's not that this mike is bad; it's just that others are better. Still, if moving-coil mikes didn't do the job, radio stations and producers wouldn't use them—and they do use them—a lot!

Ribbon Microphones

Similar in many respects to moving-coil mikes are ribbon mikes (see Figure 3.3). Some of the classic broadcasting mikes are this style. It's easy to look at some of these bulky, heavy old-timers and chuckle—until you hear them at work. They produce a mellow, warm sound that's hard to equal, let alone surpass. The ribbon mike operates on the same principle as the moving coil, except that the diaphragm and coil of wire have been replaced by a thin metallic ribbon. Sound causes the ribbon to vibrate (within the magnetic field), thus generating the current.

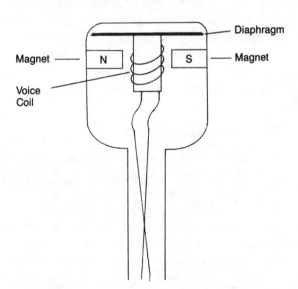

3.2 Cutaway of a moving coil mike. Sound moves the diaphragm, causing the coil to vibrate within a magnetic field, generating electrical current.

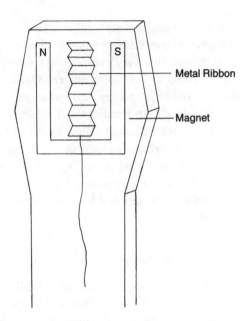

Metal Ribbon

Magnet

3.3 Cutaway of a ribbon mike. Sound causes the metallic ribbon to vibrate within the magnetic field, generating the current.

Even though ribbon mikes produce fantastic sound, they have serious drawbacks. The ribbons themselves are very fragile. A moving-coil mike can be knocked around and survive intact—not so a ribbon mike. Also, ribbon mikes are very expensive, prohibitively so for many small studios and radio stations. What's more, the sensitivity that makes them so wonderful in the studio is a real problem in out-of-studio settings. Wind, which would barely register on a moving-coil mike, might sound like a tornado on a ribbon, and performers like James Brown would probably make a ribbon mike explode.

Condenser Microphones

The *condenser mike* (sometimes called a *capacitor* or *electret mike*; see Figure 3.4) attempts to combine the positive qualities of ribbon and moving-coil mikes, while minimizing their shortcomings. The condenser is not as rugged as the moving coil but is much more durable than the ribbon. Price is also a factor: good condensers aren't cheap. They may not have the high price tag of good ribbon mikes, but they're considerably more expensive than moving-coil mikes. These mikes also have the problem which was alluded to earlier: They run on electricity provided by an outside power source. Many large consoles have what are termed *phantom power sources* built into their input modules specifically to supply needed electrical energy (*juice*) to condenser mikes. In the absence of a phantom power source, the mike has to provide its own power, frequently from batteries contained within the mike's body. This in itself isn't a problem, except that you don't want to be in the middle of an important interview or production session when the batteries run down and have to be replaced.

The condenser's sound quality is also in the middle: better than the moving coil, not as good as the ribbon. A lot of producers might dispute this last statement, and I wouldn't argue with them. Mike sound is very subjective: what sounds good to one producer might not sound so great to another. Also, how a mike sounds depends on the application: some are great for voice but only passable for instruments. Within the realm of music recording, mike choice is of paramount importance, but that choice is also (as you might guess) subjective. A piano may sound warm and mellow on one mike, sharp and crisp on another. Neither is better nor worse than the other: it all depends on what suits your ear.

The condenser microphone generates its signal in this fashion: a thin metal plate acts as a diaphragm. Positioned just behind and parallel to the plate is another plate, and between them is a small space. When sound makes the diaphragm move, the space between the plates fluctuates. There is an electrical charge running through the plates, and as the space between them varies, so does the charge they produce. Imagine two magnets. As you move the magnets closer to each other, the strength of the magnetic field between them increases. As you separate them, the field intensity diminishes. As a loud sound pushes

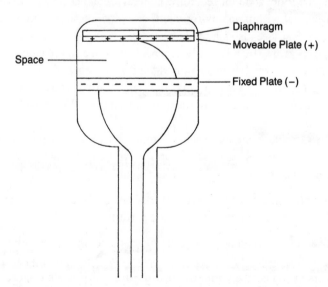

3.4 Cutaway of a condenser mike. Sound on the diaphragm alters the size of the space between the plates and results in a fluctuating flow of current.

the diaphragm plate closer to the back plate, the increased strength of the field in the space between them causes an increase in current flow from the mike. A quiet sound would also push the plates together, but not as much, resulting in less of a charge in the space, and, consequently, less current flow from the mike. An electronic component called a *capacitor* stores electricity in much the same way as the plates in a condenser mike. Hence, this type of microphone is often called a *capacitor mike.*

Microphone Response Patterns

Not all microphones pick up sound from all directions. Some do. Some can only pick up sound entering from the front. Others pick up from the front and sides, but not the back. A microphone's *response pattern* shows us the direction(s) from which a mike will pick up sound. This field around the mike can be shown on a graph called a *polar diagram* (see Figure 3.5). The polar diagram draws an imaginary circle around the mike, with the mike head at 0 (zero) degrees, and the tail at 180 degrees, showing the directions from which the mike is most and least sensitive. Also, microphones, like human ears, are more sensitive to some frequencies than to others. This may also appear on the polar diagram.

There are three basic response patterns:

1. *omnidirectional*—the mike is equally sensitive to sound coming from any direction; good for recording general *ambience* (the naturally occurring sound within an environment) or a large section of an orchestra (Figure 3.6). (See also Figure 3.5b.)
2. *bidirectional*—the mike only accepts sound in a figure-eight pattern, coming from the 0- and 180- (or 90- and 270-) degree directions; it's good for recording a two-person interview, especially if you also want to capture some (but not a lot) of the ambience (Figure 3.7). (See also Figure 3.5c.)
3. *unidirectional*—the mike accepts sound best from 0 degrees (directly in front of the mike); its good for interviewing in a noisy environment or for miking voices or instruments in a band

While omni- and bidirectional patterns are fairly consistent, there are a number of variations of the unidirectional pattern:

Cardioid—so called because of its heart shape; should be pointed directly at, or just to the side of the source; less sensitivity on the sides; rejects sound coming from the rear (Figure 3.8). (See also Figure 3.5d.)
Supercardioid—similar pattern to cardioid at 0 degrees; has slightly more sensitivity at the rear. (See also Figure 3.5e.)

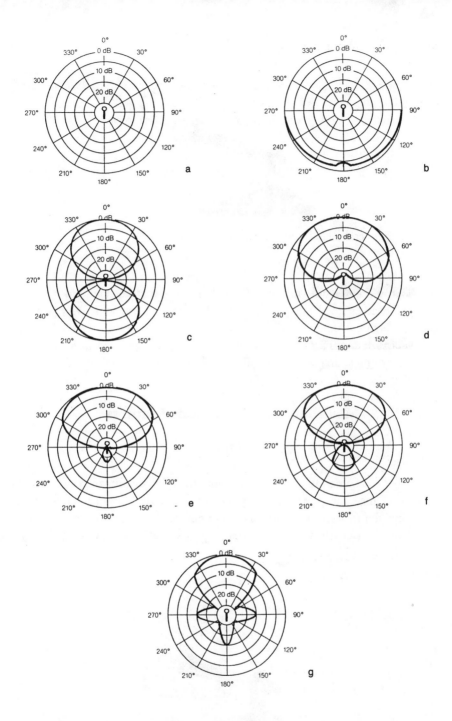

3.5 Polar diagrams. (a) Standard polar diagram; mike sensitivity is shown as a function of direction (0–360 degrees) and relative distance from the mike's head; (b) omnidirectional pattern; (c) bidirectional pattern; (d) cardioid (unidirectional) pattern; (e) supercardioid pattern; (f) hypercardioid pattern; (g) ultracardioid pattern.

3.6 Omnidirectional (or nondirectional) pattern.

Hypercardioid—narrower field of response than the cardioid at 0
degrees; good side rejection; greatly increased rear sensitivity (See
Figure 3.5f.)

Ultracardioid—extremely directional; must be pointed directly at
sound source; excellent side and rear rejection. (See Figure 3.5g.)

Miscellaneous Microphone Features

Multidirectional Mikes

Some mikes come with more than one diaphragm, thus enabling
you to change the pickup pattern. There's generally a switch on the mike,
allowing you to select the pattern of your choice. The RCA-77 model (like the
one David Letterman uses on his desk) has a switch that can make the mike
omni-, bi-, or unidirectional.

3.7 Bidirectional pattern.

3.8 Directional (cardioid) pattern.

Proximity Effect Switch

When a sound source is too close, some mikes, because of their design, have a tendency to boost the frequencies below 500 hz. This is a particular problem for announcers and singers who like to crowd the mike: their voices sound muddy. This phenomenon is called the *proximity effect*. Many mikes have a switch that allows you to slightly roll off (diminish) the bass response of the mike, thus compensating for any proximity effect. Some mikes may have a similar switch, which, instead of rolling off the bass, slightly boosts the treble frequencies. Incidentally, sometimes the proximity effect can be an asset. For example, extra bass might give a drum a little more power.

Pop Filters and Windscreens

Most mikes don't react well to sudden bursts of air. When you're close to the mike, and you pronounce a p, b, t, or k, the initial burst of air can make the mike distort, putting an annoying *pop* on the tape. A *pop filter* is a foam or felt pad inside the mike head that minimizes many of these *transient* (brief, intense) sounds. If a mike doesn't have a pop filter, you can improvise one. Bend a length of coat hanger into a circle at least 4 inches in diameter, and over this circle tightly stretch a piece of pantyhose material. Mount it between you and the mike, and you're in business. Quincy Jones used this sort of thing when he recorded "We Are the World!"

A *windscreen* is a sponge cap that fits over the mike head. Like the pop filter, it cuts down the power of transients, and, as you'd guess, it also cuts down on the roar that even a light wind can cause when it hits a mike diaphragm.

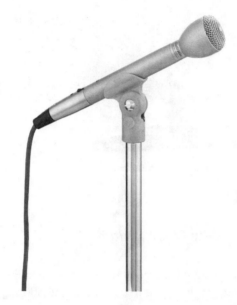

3.9 An example of a good moving-coil mike, EV 635A, rugged omnidirectional mike. The president of EV once hammered nails into a 2×4 with one of these mikes, plugged it back into the PA system, and used it for the rest of his presentation. (Courtesy of Electro-Voice.)

3.10 EV RE-20. Probably the most popular on-air mike around today. Cardioid response pattern. Note the ports along the mike's body which cancel sound entering from the rear and sides. A good clone, the RE-15, is also available. (Courtesy of Electro-Voice.)

Choosing the Right Mike

The art of using the right mike at the right time can only be mastered over time. Entire books have been written on the subject (see the "Suggested Bibliography" at the end of the book), so don't think this brief overview provides all there is to learn. Figures 3.9–3.13 give you an idea of the range of styles of mikes available today.

Microphone Technique: Positioning

Learning how to use a microphone correctly involves two elements: vocal technique, which is covered in Chapter 8, and mike positioning, which is covered now. The two key aspects of mike positioning are your *distance* from the mike and your *focus* in relation to the mike.

1. *Distance.* The tighter (closer) you are to your mike, the better. Obviously if you're some distance away from the mike (a foot or more), the level will be

3.11 RE-18 is one of a series of hypercardioid mikes. Unlike the 635A, 50, and 20, which are primarily voice mikes, this is great for miking music. (Courtesy of Electro-Voice.)

3.12 Two classic ribbon mikes, the RCA-77 DX and the legendary RCA-44. Don't let their clunky appearance fool you. Even though these two warhorses have been knocked around, they still sound great! The 44 is bidirectional. The 77-DX has omni-, bi-, and unidirectional capability. (Courtesy of RCA.)

extremely low, and you'll have to compensate by cranking the mike pot up, which, in turn also cranks up the noise. Stay within 1–2 inches of your mike, but don't jam your face against the mike. Bumping the mike is not only amateurish and noisy, but it's also probably unsanitary. Don't touch—ever.

2. *Focus.* If you're less than 6–8 inches from the mike, never speak directly into the front of the mike head, or you'll risk popping many consonants. We have a number of explosive consonants in our language (b, d, k, p, t), which are formed by releasing built-up pressure within the mouth. These small bursts of air may not feel like much to you—but they can wreak havoc on a sensitive mike diaphragm. Recite "Peter Piper picked a peck of pickled peppers," directly into a *tight* mike, perhaps 1 inch from your mouth. When you play back, the *p*'s will sound like small explosions. Popped consonants are the sure sign of a novice, and many beginners wrongly assume that the problem lies in their pronunciation. So they try to deemphasize the problem consonants, only making their reading or speaking sound unnatural and stilted.

The secret to eliminating popping consonants lies in the mike's position. If the head of the mike is pointed straight at your mouth, but slightly off to the side, you'll solve the problem. To determine how far off to the side to position

3.13 A condenser microphone. C-48, with omni-, bi-, and unidirectional capability. Bidirectional mikes or mikes with bidirectional capability often, though not always, have flat sides, as opposed to a round mike head. (Courtesy of SONY Corporation.)

the mike, try this: imagine a long tube (like the one from the center of a roll of paper towels) projecting from your mouth. As long as you keep your mike outside that tube, you won't pop. Conversely, if any part of the mike head is inside the tube, pops are almost guaranteed. Also, don't wander. Like most announcers, you may make manual and facial gestures while reading. This is perfectly natural. Just make sure that your movement doesn't accidentally carry the imaginary tube too close to the mike head.

Experiment with mike positioning. Shifting your mike even a little can noticeably alter how it colors your voice. Eventually, you'll automatically pull the mike into the correct position as easily as you pull on a pair of comfortable sneakers.

Review

1. A microphone is a *transducer*, a device that changes energy from one form into another.
2. Mikes differ in the way they generate current and in their response patterns.
3. The three types of mikes are the moving coil (or dynamic), the ribbon, and the condenser (or electret).
4. Assessing mike quality is completely subjective. Moving-coil mikes are particularly popular in radio stations because of their durability, affordability, and generally good sound.
5. The three basic response patterns are omnidirectional, bidirectional, and unidirectional (cardioid). There are a number of variations of the unidirectional pattern.
6. Some mikes have a switch to minimize the *proximity effect,* an annoying boost in bass frequencies, due to the announcer being too close to the mike.
7. Pop filters and windscreens cut down on unwanted environmental noise and popping consonants.
8. Position the mike fairly close to your mouth, but slightly off to the side, to eliminate popping consonants.

4

Tape and
Tape Recorders

If the console is the brain of your production system, your tape recorders (reel-to-reel, cassette, and cartridge) are your most important tools. They are to a radio producer what brushes and canvases are to an artist: they provide the medium on which intangible ideas become physical reality.

Audiotape

Self-proclaimed prognosticators tell me that in the future, tape will be obsolete, that recording will be done on special discs or fibers or some other as-yet-unknown medium. They may be right, but for now, as it has been for over 30 years, tape is the name of the game.

One of the few good things to come out of World War II was the concept of electronic recording. The Germans, unbeknownst to us, had developed a machine through which sound (actually an electronic representation of sound) could be recorded onto a wire and then played back. American technology took the idea a step further by replacing the wire with a narrow, flat ribbon made of acetate and then coating one side of the ribbon with microscopic particles of iron oxide. The recording process, depended on these metal fragments being moved around by a magnetic field.

Several factors affect audiotape quality:

- acetate versus mylar stock
- type of coating
- tape width
- tape thickness

Today, *acetate stock* has largely been replaced by *mylar*, which is stronger, more durable, cheaper, and more resistant to changes in temperature and humidity. However acetate is still around because it has one advantage over mylar—acetate won't stretch. When mylar tape stretches, as it will under sufficient lateral (sideways) stress, it becomes useless. You can't record on it, and anything already recorded on it is ruined. Acetate, when subjected to the same stress, breaks cleanly, and can be easily spliced together. This characteristic alone accounts for the fact that acetate tape hasn't gone the way of the dinosaur.

The *coatings* used by tape manufacturers have changed over the years, all geared toward allowing you to record at a louder level without distortion, thus increasing the signal-to-noise ratio. Today, the more expensive tape has chromium dioxide or pure metal (no oxide) coating. Many of these tapes are labeled *high bias*, referring to the electrical current the tape recorder lays down to enhance the recording process. If you're using a high-bias tape, your tape recorder may require bias adjustment. Check with your engineer. If you're lucky, your decks may have simple bias-select controls, just as many cassette decks today have chromium and metal tape settings.

Tape *width* ranges from the ⅛-inch tape used in cassettes, to the 2-inch tape used with large, multichannel recorders. Tape width is an important variable because of a pivotal principle:

> The more information you can put on the tape, the more similar the recording will be to the original sound.

That's what *fidelity* is, and it's of paramount importance to a producer. It's analogous to the tiny pixels (cells, dots, short for *picture elements*) that make up your TV screen. The more pixels per square inch, the clearer and better defined the picture will be. The more information you can put on a length of recording tape, the clearer, cleaner, and more accurate the recording will be. Obviously, a length of ¼-inch tape can hold only 25% of the material that an identical length of 1″ tape can hold. However, due to the great improvement in tape and tape recorders over the years, most radio stations and small recording studios find ¼-inch tape perfectly adequate for their needs. As a result, most reel-to-reel recorders made today are built to accommodate ¼-inch tape.

The next factor to consider is the tape's *thickness*. Recording tape is measured in mils (1 mil = 1/1,000 -inch), with most tape today measuring 0.5 mil, 1.0 mil, or 1.5 mil thick. A 7-inch reel will hold 3600 feet of 0.5-mil tape. With 1.5-mil tape, the reel will only hold 1200 feet. Although it would appear that the 0.5-mil tape is the better bargain, assuming comparable prices, don't be fooled. Remember that mylar tape stretches; and the thinner the tape, the more likely it is to stretch.

One other problem with thin tape is *print-through*: recorded tape contains a magnetic signal. If the tape on a reel is too thin, a strong signal on one layer can actually affect the signal on the layer of tape directly below it. It's better to buy 1200 feet of usable tape than 3600 feet of garbage. *Never* use anything thinner than 1.5-mil tape!

4.1 Standard tape reels, 10½, 7 and 5 inches in diameter.

In general, don't buy cheap tape. Use a name brand, or ask someone who's opinion you respect to recommend a brand. If you're not sure about a particular make of tape, buy a single reel, and try it out on some noncritical production. The oxide from brand X tape can also flake off as it passes through your equipment. Unless you're the type of person who would build a house on quicksand, stick with a reputable product.

In addition to considering the tape quality, you must consider what size of tape *reel* to use. Most of the tape reels used today are 5, 7, or 10½ inches in diameter (see Figure 4.1). I like to use a larger reel (or a smaller reel with a lot of tape on it) for a couple of reasons. First, the added weight means the reel is more stable and more likely to keep the tape on a straight path. Second, the reel is less likely to run out of tape while you're in the middle of a session.

Tape Recorders

Although radio stations and production studios utilize three tape formats—reel-to-reel, cassette, and cartridge—an analysis of the reel-to-reel recorder will serve to explain much of what you need to know about the other two.

Like consoles, tape recorders range from the simple to the complex, from the somewhat expensive to the extremely expensive. Also, like consoles, all tape recorders are alike in that they are all equipped with the same basic systems:

1. a transport system to move the tape
2. a record/playback system

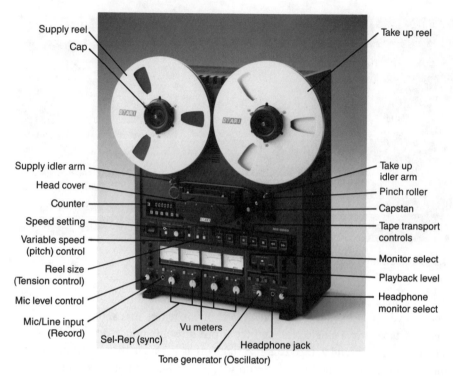

4.2 Typical, professional 2-track tape deck. Most of these features are standard on modern units. (Photos courtesy of Otari Corp.)

Tape Transport

The tape transport system is responsible for moving the tape smoothly from reel to reel. The key word is *smoothly*, because erratic tape movement will produce an erratic recording. First, let's identify some components along the tape path (see Figure 4.2):

Supply (or **flange) reel**—the start of the transport system; feeds tape into the recording system

Supply idler—designed to help keep the tape's tension uniform; can be an arm or a rotating cylinder

Tape guides—designed to keep the tape in a consistent path; guides can be fixed or movable

Take-up idler—same as supply idler; also engages capstan and reel motors (see the following paragraph)

Scrape filters—small metal cylinders positioned between the heads (discussed in the next section); help maintain tape tension

Lifters—lift the tape off the heads (the devices that do the actual recording and playing) during REWIND or FAST FORWARD movement; prevents head wear

Takeup reel—the end of the transport system; receives and collects tape

The heart of any tape transport system (see Figure 4.2), is the *capstan/pinch roller assembly*. The *capstan* is a small drive shaft coming from a motor inside the recorder. When the motor is engaged, the capstan rotates in a clockwise direction. Frequently the capstan motor is tied to the take-up idler: when the idler is lifted during tape threading, the capstan motor automatically engages. If the idler were to drop, as it would if the tape supply ran out, the capstan motor would stop. In some machines, the capstan is activated by a photoelectric cell somewhere in the tape path. When the tape is threaded, the beam is broken and the capstan rotates.

The *pinch roller* is a floating wheel, usually positioned just above (occasionally just below) and mounted parallel to the capstan. The tape passes between the capstan and pinch roller; and when the recorder is switched into play or record mode, the pinch roller (formerly made of rubber, now made of neoprene or a similar material) clamps down against the capstan, pinning the tape. Because the capstan is rotating clockwise, it imparts to the floating pinch roller a counterclockwise motion; and the tape, caught between them, is squeezed out toward the take-up reel. If the pinch roller is mounted *beneath* the capstan, the capstan will turn counterclockwise, thus causing the pinch roller to turn clockwise. In either case, the tape is pulled to the right.

Transport Controls

How these controls move the tape should be familiar. (Figure 4.2)

Play—moves the tape from left to right; tape is on the heads

Record—same movement as *play*

Rewind—disengages capstan and pinch roller; supply reel rotates rapidly clockwise; tape is lifted off the heads

Fast forward—disengages capstan and pinch roller; take-up reel rotates rapidly counterclockwise; tape is lifted off the heads

Stop—self-explanatory; tape is lifted off the heads

Edit—disengages the take-up reel without affecting the rest of the transport system; lets you spool off unwanted tape during editing

Threading the Tape

Your tape recorder should be equipped with a couple of caps to hold the reels on the spindles. If a recorder is mounted flat (parallel to the floor), a lot of people don't use the caps, figuring that gravity will hold the reels in place. It's a judgment call, but if the deck is mounted vertically (upright), there's no question: *always* use the caps. It's difficult to impress a client with your expertise when a reel comes flying off your tape recorder.

The threading procedure with this type of transport system (called an *open loop system*) is as follows: the tape comes off the left side of the supply reel, through (and/or around) the supply idler/guide, through the tape guides, across the tape heads, through (and/or around) the take-up idler/guide, and onto the right side of the take-up reel. This may sound silly, but make sure the recording side of the tape is facing the heads. Generally, the oxide side and the back are different colors or different finishes (one side dull, the other side shiny.) The colors and finishes used vary from manufacturer to manufacturer. The tape I use has a dull black back, and a shiny brown oxide side. You can't record on the back of the tape.

You may run across another transport system, called a *closed-loop system*. The main difference here is that the tape is squeezed between the capstan and *two* pinch rollers. Producers who use machines with this system say that the tape gets better support going across the heads. Personally, I don't see the closed-loop design as being any better or worse than the open-loop design: it's just that the open-loop format has always been the standard. At any rate, it's your choice.

Reel Motors

In professional tape machines, there are a number of motors, one controlling the capstan, and others controlling the supply and take-up spindles. You can see what these motors do by turning the deck on, manually lifting the take-up idler, which engages the motors, and running through the transport controls. In the *play* mode, the supply spindle will begin spinning rapidly clockwise, while the take-up spindle spins counterclockwise. The amount of *torque* (turning power) produced by the spindles is great enough to keep tape taut on both sides of the capstan, but small enough to keep from stretching the tape. In the rewind mode, only the supply spindle turns; in fast forward mode, only the take-up reel does. Reel size control makes the reel motors compensate for differences in torque between reels of different sizes.

Tape Speed Control

All good decks have two (or more) speed settings. The most frequently used speeds are 7-½ and 15 inches per second (ips), with large multi-channel decks capable of running at 30 ips. The speed of the tape is critical to the recording process. *The faster the tape moves during recording, the better the fidelity.* The reason for this fact is that a greater amount of tape passing through the system per second can hold more information. One second of tape traveling at 7-½ ips is only half as long, and, consequently, can only hold half the magnetic information as one second of tape travelling at 15 ips. The more information the tape can hold, the better the frequency response and fidelity. This same principle applies to tape width. Until the advent of special noise-reduction devices, cassette fidelity was terrible because of the tape's narrow width (⅛-inch) and very slow speed (1-⅞ ips).

For most broadcast applications, 7-½ ips is the generally accepted standard. Today's tape recorders are good enough so that recordings at 7-½ ips sound fine, especially if you're just recording a speaking voice. With music production, however, because fidelity is more of a concern, never record at slower than 15 ips.

Variable Speed Control

Some tape machines come with a variable speed oscillator, which allows you to slightly speed up or slow down the tape movement. This can be used for special effects (see Chapter 11, "Tricks of the Trade") or to change the length of a piece of production. For example, a commercial that's :31 long can be shrunk down to :30 with your speed control (:30 and :60 are the standard broadcast lengths for commercials). Watch out! Altering the speed of the tape also alters the pitch of the recording. As a general rule, you can safely pick up 1 or 2 seconds for each thirty seconds of material without the pitch change becoming too noticeable. If you try to pick up more than 2 seconds, you'll have problems. If you speed up a :35 taped spot to make it a :30, the announcer will sound like he's been inhaling helium.

The Record/Playback System

The Tape Heads

Just as the capstan and pinch roller are the keys to the transport system, the *tape heads* are the keys to recording and playing back sound. The standard tape recorder has three tape heads (see Figure 4.3), which are almost always arranged in this order from left to right:

erase head   record head   playback head

Their names tell you what they do: erase a signal from the tape, record a new signal onto the tape, and play back an existing signal from the tape, respectively. Smaller reel-to-reel machines, as well as cassette players may have only two heads, one erase head, and one head that functions as a combination record/playback head. Some professional models come with an extra playback head. If you're not sure which head on your machine does what, check the owner's manual.

The heads can be made of glass, metal, or a combination of both; each head houses a tiny electromagnet, which can move (or be moved by) the metal particles

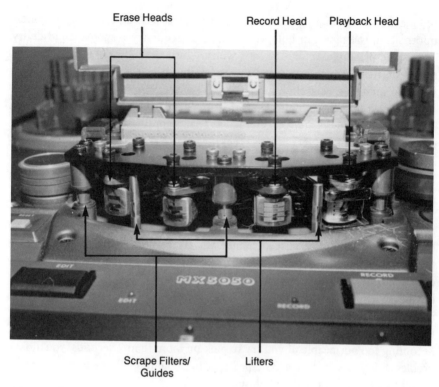

4.3 A peek under the head cover. This deck has two erase heads to ensure more thorough erasure. (Photos courtesy of Otari Corp.)

on any tape passing by. Also, these tiny metal particles actually can wear down the heads over time. That's why the tape lifters keep the tape off the heads during rewind or fast-forward modes—because the damage would be the greatest in these modes.

When the machine is put into the record mode, a number of things happen:

1. The *erase head* is turned on, and magnetically moves the particles into completely random patterns, breaking up any organized patterns created by a record head. These organized patterns provide the key to recording. If the patterns are made random, there will be no sound recorded on the tape. (Purists will say that there's really no sound on a tape, any more than there is sound in the grooves of a record. They're right, but it's easier for me to refer to the sound on the tape rather than elecromagnetically induced patterns.) By neutralizing any existing patterns, the head, in effect, erases the tape.

2. When a sound is changed via a transducer (like a mike) into an electrical current and sent to the *record head*, fluctuations in the current (representing

fluctuations in the original sound wave) cause the head to produce a fluctuating *magnetic field* which pulls the tape particles into organized patterns. Follow the path: the sound is converted (via the mike) to *electricity*, which is converted (via the head) into a *magnetic field*, which is converted into *particle patterns* on the tape. In addition, the particles require a little kick to overcome their inertia, so the record head also produces an extremely high-frequency signal called a *bias signal*. The *bias signal* forces the particles to align themselves with the magnetic field coming from the head, sort of like your mother forcing you to eat your broccoli. So, the tape then leaves the record head imprinted with an electromagnetically generated representation of the original sound.

3. At the *playback head*, the recording process is reversed when the machine is put into the play mode. The particles on the tape alter the magnetic field in the playback head, which causes the head to generate a correspondingly fluctuating electrical current, which is sent out of the tape recorder and eventually (via a speaker, for example) converted back into the original sound.

This, in a nutshell, is how a tape recorder works.

Miscellaneous Record/Playback Controls

Several controls allow you to regulate the tape recorder's actions:

Input control—sets the deck's Vu meters in response to the incoming signal. The deck should accept both mic and line (low and high) levels, and may have separate controls for each.

Output control—adjusts the level coming from the tape; if the deck doesn't have this feature (and some don't), then the output level is set only by the console.

Monitor controls—determine what the deck meters will show—either (a) *input* (or *source* or *record*), the level coming into the deck; or (b) *repro* (or *tape* or *playback*), the signal coming from the tape.

Record select (or record set)—selects the tape track(s) on which the recording process is to take place.

The concept of *tape tracks* (see Figure 4.4) is crucial to creative production. Although the recording surface of audiotape appears uniform, the tape recorder, by means of the record and playback heads, divides the tape electronically into horizontal sections called *tracks*, and it can record different material on each track. If you look at your record and playback heads, you'll see not only the track markings, but also spaces between the tracks, called *guard bands* (see Figure 4.5), which are designed to minimize the possibility of a signal leaking onto an adjacent track. The width of the heads determines the number of tracks. There are full-track heads less than an inch high, which record only a single

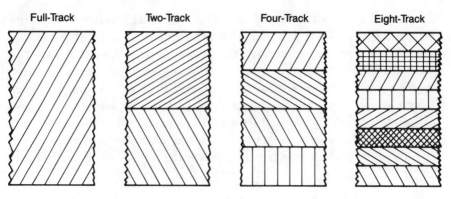

| Full-Track | Two-Track | Four-Track | Eight-Track |

4.4 Tape tracks.

track, filling the entire tape width; and there are heads more than 2-inches high, which can record up to 32 tracks. If your decks use ¼-inch-wide tape, the tape heads may be anything from single (full) track to four-track.

Remember the relationship between sound fidelity and tape width? The same holds true for *track* width. The narrower the track, the poorer the sound quality. That's why the tape must be wider if it's to accommodate a large number of tracks.

Tape Recorder Operation

Preparation

Before you begin operating any tape machine (reel-to-reel, cassette, or cartridge), three items of business *must* be taken care of:

1. Set the levels, aligning the Vu meters with those on the console (see Chapter 2 for the procedure).

2. Clean the heads (see Figure 4.5). Use long cotton swabs and isopropyl alcohol, both of which you can buy in bulk from any drugstore. Soak the swab in alcohol and, applying a little pressure, clean every surface the recording side of the tape comes in contact with—heads, guides, pinch roller, everything. While you're at it, clean the capstan. Lift the take-up idler so the capstan will rotate, hit play, and then the capstan and pinch roller while they spin. Keeping the heads clean is of paramount importance. There's nothing more frustrating than recording what you think is a good track, only to have it sound muddy on playback because of a little dirt on one of the heads. Also, because the tracks on the heads are so narrow, it takes only a speck of dirt to undo your work.

In fact, if I'm working on a machine for a protracted length of time, I'll clean the heads at least every hour—better safe than sorry.

3. Erase your tape. Don't assume that your erase head will do this job for you. Use a bulk eraser (often referred to as a *bulker*) instead. This handy tool (see Figure 4.6) is merely a good-sized electromagnet, which, like your erase head, obliterates any existing particle patterns on the tape, thus erasing it. However, the bulk eraser is much stronger, so the erasure is more thorough.

There are a couple of things you should know about using a bulker. First, drawing *each side* of the reel, cart or whatever slowly across (or under if you've got a hand-held model) the bulker just once should do the trick. Remember that you're using a *very* strong magnet, so there's no need to spend 5 minutes scrubbing the tape; 10 seconds is plenty.

Second, after the tape is erased, but before you shut the bulker off, move the tape at least an arm's length away from the bulker. Why? Because when the magnetic field is shut off and the tape is within the field, the receding field can actually put noise on your tape. Have you ever run what you thought was a clean tape in fast-forward mode and heard a rapid whup-whup-whup coming from the speakers? Those whups are like footprints left behind by your eraser. To avoid these, pull the tape out of the field before switching the eraser off.

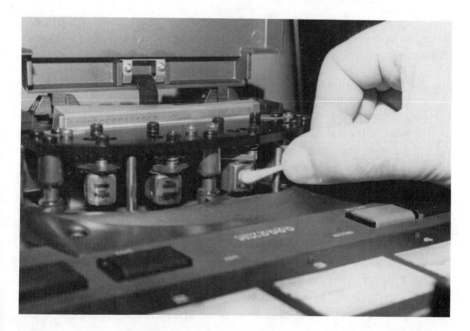

4.5 Tape preparation. Before beginning work, clean the heads, tape guides, pinch roller/capstan or anything that will contact or otherwise affect the oxide side of the tape.

4.6 A hand-held bulk tape eraser. This one's good for video- as well as audiotape. (Reproduced with permission from the Tandy Corporation.)

Recording

The following recording procedure is typical. There will be some variation from deck to deck, but most equipment will have a recording procedure analogous to this one.

1. Set the monitor control to input, to set your level. Technically, this step isn't mandatory. The deck will record with the monitor set to repro; however, with repro/playback mode, the meters will be showing you the level *already recorded* on the tape, and not the level *going onto* the tape. If all your levels have been set properly prior to recording, the setting of the tape monitor control is not particularly important.
2. Select the track onto which you want to record, using the record select control.
3. Push the play and the record controls. On some decks, these need to be pushed simultaneously.

Steps 2 and 3 must be followed, or the machine will not record. Even if you push the play and record controls, if a track hasn't been selected, no recording will take place. Because you need to use two controls (the record select and the play/record controls), the chances of accidental recording are minute.

Note: Keep the tape recorder pots on the console *closed* during recording to avoid getting *feedback* (if the tape monitor is set to input) or *echo* (if the monitor is set to repro).

Playback

The following procedure for playing works for almost all tape recorders.

1. Set the monitor select control to repro (or *tape* or *play*).
2. Turn off the record select.
3. Open the appropriate pot on the console.
4. Rewind the tape, stop it, and then hit play.

To help you find the beginning of your recording, your deck may come with a cue feature. This keeps the tape up against the heads even when running the tape in rewind or fast-forward modes. The cue control is a mixed blessing: it makes finding a specific point on the tape much easier, but it also wears down the heads, especially if you do a lot of fast tape movement with the cue control on. Use cue sparingly: heads are *very* expensive.

Synchronization

Your tape recorder's synchronization control may be labeled *sync* or *sel-sync* (selective synchronization) or *sel-rep* (selective reproduction). By any name, it gives the producer the crucial ability to record new material onto a new track and to synchronize it with preexisting material on another track. With synchronization, a singer can record a melody on one track, rewind the tape, and sing harmony on another track. A sound effect that doesn't fall exactly where it should in a recording can be rerecorded on another track, without having to redo the other tracks. If a band has a recording session, and one of the musicians can't attend, those in attendance can lay down their tracks, and the absent member can lay down the remaining track at another time.

Try the following experiment to see what the sync control does.

1. On Track 1 (or on the left track if you have a simple stereo deck), record your voice, counting slowly from one through ten.
2. Rewind the tape.
3. Set the tape monitor control to repro (or tape or playback), so you can hear what you've recorded.
4. While listening to your first recording (through headphones), record another ten count, but on Track 2 (or the right track). Try to keep the two counts simultaneous.
5. Rewind the tape, and play the two tracks back together.

You might expect two simultaneous counts, but surprisingly, you'll hear the second track lagging significantly behind the first. The reason lies in the

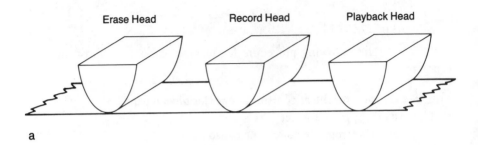

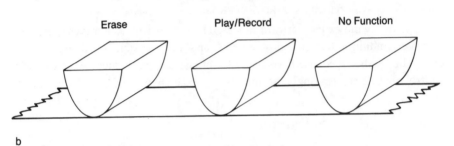

4.7 Head functions in (a) play (repro) and (b) sync (sel-rep) modes.

placement of the heads. When you listened to the first track playing back, what you heard was the tape being read at the playback head, but when you recorded your second track, the signal going through your mike was imprinted onto that second track at the record head. The space between the two heads accounts for the gap between the two recordings (see Figure 4.7).

The sync feature of your deck was designed to eliminate the problem. Earlier, we learned that although the record and the playback heads do different jobs, they're actually quite similar in construction, so much so that in many cassette decks, a single head performs both functions, record and playback. When you engage the sync mechanism, the record head is changed to a combined record/playback head. Now, try repeating the counting experiment, but set the monitor select to sync instead of repro when you put down the second track. This time you'll hear the first track *and* record the second track both at the record head. Play the two tracks back with the monitor select set to repro, and the two tracks should line up perfectly.

If you want to hear exactly what the sync feature does, while a tape is playing (a voice recording would be best), change the monitor select from repro to sync. You'll hear a noticeable jump, because when the deck switches playing from the playback to the record head, the small section of tape between the heads is bypassed—so the recording jumps. If you keep the tape rolling, and switch from sync back to the repro mode, you'll hear a bit of the tape repeat itself. Why? Because when you switch from sync to repro mode, the bit of tape between the heads that was just played by the record head now passes over the playback head, and you hear it again.

As you'll see when we cover some of the tricks you can perform using synchronization, this single feature makes much of today's music production possible. The pioneering work that was put on tape in the late 1940s and early 1950s by people like Les Paul and Mary Ford and Patti Page had and continues to have a profound influence on musicians and producers today.

Other Tape Formats

Besides reel-to-reel recorders, today's radio production studios invariably have cassette and cartridge tape recorders.

Cassettes are a good example of good news/bad news. The good news is that for convenience, cassettes are hard to beat; the fact that cassettes and cassette player/recorders are generally small, portable, and easy to use makes them ideal for home and studio use. When a client wants to hear a commercial, even though the whole company is at a conference in Pago Pago, popping a cassette into the mail is a snap. Since its appearance in 1964, the cassette's progressive rise in popularity has been matched by a decline in album (LP, for long playing) sales. Today, in 1991, the LP accounts for less than 10% of recorded music sales and, according to many industry watchers, may be headed for extinction. People frequently point to the CD as the culprit; but the real damage was done over the past 15 years by the cassette.

The bad news about cassettes is their quality. Because of its narrow tape width (⅛-inch) and slow speed (1⅞-ips), early cassettes had a signal-to-noise ratio that was nothing short of pathetic. Also, longer-playing cassettes use tape that's only 0.5 mil thick, and it's not uncommon for the tape to become stretched and tangled inside the machine. As mentioned earlier, the sound quality problem has been eased (but not eliminated) with the advent of noise-reduction systems like Dolby and dBx (see Chapter 13, "Processing"), and high-output tape, which allows for recording at higher levels without distortion, thus masking the noise. Aside from the sound quality, a cassette's playback value in the studio is minuscule. Cuing a cassette is tough, so if your production requires split-second timing a reel is infinitely more reliable; and editing a cassette is next to impossible on standard equipment.

Cassette recorders operate like miniature reel-to-reel machines, with the tape traveling between two hubs via a capstan and pinch-roller assembly. Tape guides are found inside the cassette shell itself.

Tape cartridges (or carts) and cart machines came on the scene around 1960; to the average announcer, they represented the greatest advance in the radio industry since the invention of the transistor. The *broadcast cartridge* (not to be confused with the 8-track cartridge, the Edsel of the music industry; see Figure 4.8) consists of a loop of tape wound around a single hub. Like the cassette, the tape guides are inside the cartridge. What made the cart a godsend was its convenience. Previously, recorded material not on disc was on reel; if you had a lot of commercials to run, that meant a lot of threading, rewinding, and hunting for the start of the spot. With the cart, that all changed. When the cart machine is put into the record mode, a low-frequency tone pulse is put on the tape at the head of the recording. This tone (called a cue tone) occupies a track that does not play over the air. The material is recorded on the tape *immediately* after the tone. When the recording is completed, the tape loop continues to run, and when the cue tone comes around (and just behind it, the start of the spot) and hits the playback head, a sensor stops the cart. The cart automatically recues itself to the head of the spot. The next time the spot is to air, the announcer pops the cart into machine, hits the play button, and the spot's right there.

Because carts use standard ¼-inch tape and run at 7-½ ips, the sound quality is quite good; so much so that by the early 1970s, more and more stations were using carts not only for commercial and promo material, but also for music. Today, this situation is almost universal. Record a song on a cart, and you eliminate the problems that discs are heir to—skips, scratches, pops, accidental bumps of the turntable, etc. The cart will provide more plays than the disc, and if something happens to the cart, the disc is still nice and clean and can be rerecorded onto another cart.

If the tape in the cart wears out, you can repack it with new tape or buy another cart. Repacking is a pain. It requires special lubricated tape (so the loop will run smoothly), a tape winder, and a healthy dose of patience. However, with the rising cost of carts, especially the better stereo ones, repacking is a workable, albeit annoying option.

The only other glitch you may run into with carts is with the pressure pads. These are small sponge pads that are mounted inside the cart and press the tape outward against the heads. Eventually, as the pads lose their resilience and the tape isn't pushed firmly enough against the heads, any recording on the tape will sound muddy. Not all carts use pressure pads: many rely on the cart tape guides and tension to keep the tape flush, but if your pads aren't springy to the touch, replace them. Unscrew the cart cover, work the old pads out, put new pads in, and secure them with a drop of airplane glue. In the mid-1980s, the price of pads skyrocketed, from a nickel apiece to around 50 cents each. The cart manufacturers obviously wanted you to buy new carts rather than fix up your old ones.

In most radio stations, production is generally done on a reel and then dubbed from the reel onto a cart. Cart machines are designed for ease of operation,

4.8 A tape cartridge, the radio announcer's best friend. (Photo: Permission of Audiopak, Inc.)

and it takes a concerted effort to mess things up. However, cart recording is susceptible to a couple of glitches that are easily avoided:

1. Always make sure your cart is erased before you record onto it. Cart machines aren't really good at erasing. So if you try to record new sound over old, you're likely to get what we refer to as a *double header.*

2. *Never* record over a splice. The recording will momentarily drop out, and so will your standing among radio producers. There are devices on the market called "splice-finders," which run a cart at rapid speed, and stop when the splice passes between the capstan and the pinch roller so your recording starts just after the split. Many of these machines also erase the cart at the same time. As expensive as a splice-finder is, it's well worth the investment. If you've ever tried to eyeball a long cart watching for the splice to come around, you know how much of a boon a splice-finder can be. Get one!

Carts and cassettes, as you might guess, have similarities and differences. The transport systems both work on a capstan/pinch-roller assembly, but each assembly is slightly different. In the cassette player, the capstan is inserted into the cassette behind the tape, and the pinch roller clamps against the tape from the outside. A cart, on the other hand, has a large hole in the bottom of the shell, through which the pinch roller rises in an arc, clamping the tape outward against the capstan, which is mounted vertically in the rear of the machine. You can't really edit on either of these tape formats, but the cart's cuing feature makes it much more useful in the production studio; most radio production is mixed onto cart, but a cart and cart player can't travel as easily out of the studio as a cassette and cassette player can.

Even with the advances in sound technology (CDs [compact discs], digital processing, etc.) that have taken place over the past few years, tape is still our primary recording medium. This situation may (and probably will) change. Nonetheless, in this business, which can change nearly overnight, tape has been the mainstay for nearly 40 years. That says a lot.

Review

1. Today's recording tape consists of a mylar ribbon with a metallic coating. Tape width ranges from ⅛- (cassette tape) to 2-inches (professional multitrack tape). Tape thickness ranges from 0.5 to 1.5 mil. Thin tape is to be avoided because it's prone to stretching and to print-through. Standard broadcast production tape is ¼-inch wide, and 1.5 mil thick.
2. More information imprinted on the tape will result in greater fidelity. As a result, wider tape and tape tracks, as well as faster tape speed, are desired. Most radio production is done at either 7-½ or 15 ips (inches per second) speed.
3. A tape recorder consists of both a *transport system*, by which the tape is moved, and a *record/playback system*, by which signals are imprinted and retrieved.
4. The capstan and the pinch roller are the heart of the transport system.
5. The reel motors keep the tape taut in the play and record modes, and they move the tape in the fast-forward and rewind modes.
6. Most professional reel-to-reel decks have at least three heads—erase, record, and playback.
7. The erase head randomizes the metal particles on the tape, thus erasing any previous imprint.
8. The record head pulls the metal particles into patterns that are analogous to the original signal.
9. The playback head reads the patterns and reconstructs the original sound.
10. Recording tape is divided by the recorder into longitudinal tracks. Different signals can be imprinted on each track. The number of tracks depends on the tape width and the size of the heads. Heads range from small, single-track heads, to large (2-inch high) 32-track heads.
11. Before beginning work, align all Vu meters, and clean the tape heads and all surfaces that contact the oxide side of the tape.
12. The sync, sel-sync, or sel-rep feature allows you to synchronize new sound with sound already recorded on the tape by allowing you to

monitor the existing track at the record head instead of at the play-back head, thus nullifying any delay.

13. Cassettes are convenient, but because of narrow tape width and slow tape speed (1-⅞ ips), the sound wasn't very good until the advent of noise-reduction systems, like Dolby. Furthermore, because the tape is thin (0.5–1.0 mil), it often stretches.

14. Broadcast-tape cartridges are indispensable to the modern radio operation because of their convenience, their self-recuing ability, and their high-quality sound (¼-inch wide tape, 7-½ ips speed).

5

Turntables

There's no denying the fact that CD (compact disc) players are appearing in more and more production studios. It's also true that CDs offer decided advantages over standard discs, especially in the areas of durability and overall sound quality. Still, you can exercise necessary manual control over a record that's impossible with a CD. Also, there are a lot of tricks and special effects you can perform only with a turntable and a record. Sound effects and production-music libraries are usually issued on discs. CDs are great, but they don't (as yet) offer the flexibility and control of a disc and a turntable.

As is the case with tape recorders, a turntable consists of two systems, a drive system (to move the turntable platter) and an electronic reproduction system (to reproduce sound from the disc).

Drive Systems

Early turntables were rotated by a *belt drive system* (see Figure 5.1). The turntable motor turned a wheel, which acted as a pulley, and a belt connected the pulley wheel to a shaft underneath the center of the *platter* (on which the discs rested). The belt could be shifted onto wheels of various diameters to run the turntable at different speeds. Obviously the weak link in the system was the belt. It could stretch, in which case the turntable would not run up to speed. The belt also could break, in which case the turntable wouldn't run at all, or, if the machine sat unused for a long time, the belt could develop a bump where it wrapped around the pulley, making smooth operation impossible. The only requirement of the platter is to rotate at a *consistent* speed *all* the time. With a belt-driven system, uniformity could not be guaranteed. There are turntables made today with this drive system, and some producers like them because they feel the belt-drive is less noisy than other systems. I don't agree, but you should try all of these drive systems, and make your own decision.

The next advance in turntable technology is the *rim-drive* or *idler-wheel-drive system*, which is still in widespread use today (see Figure 5.2). The rubber belt is replaced by a free-floating rubber (or rubberlike plastic) disc (the idler wheel), mounted under the platter near the rim. A drive shaft protrudes from the motor near the idler wheel, and when the machine is put into gear, the idler wheel wedges itself tightly between the drive shaft and the rim. The drive shaft thus imparts rotary motion to the idler wheel, which, in turn, imparts rotary motion to the rim, thus rotating the platter. The drive shaft is stepped, so that even though the shaft itself rotates at a single speed, the circumferences of the different segments of the shaft will move at different speeds. Positioning the idler wheel at different points along the shaft will cause the wheel and, consequently, the platter to turn at different speeds. Many of these turntables have a lever that changes the turntable's speed by shifting the position of the idler wheel up or down the drive shaft. The lever should also have a neutral position. When the turntable's not in use, it's a good idea to leave it in the neutral position. If the idler wheel is left wedged between the drive shaft and the rim for an extended time, it could develop a flat spot, which does for your sound reproduction what a flat spot on a tire would do for a ride in your car. All things considered, the idler-wheel drive is far superior to the belt-driven branch of the family. The idler wheel is very durable. Formerly made of rubber, which could, over time, soften and slip, idler wheels are now made of neoprene, a rubbery plastic that is also used to make pinch rollers on many tape recorders.

5.1 An old belt-drive turntable with an external belt. Note the stepped capstan. The higher on the capstan the belt rides, the slower the platter turns.

5.2 Rim-drive turntable. The spinning capstan rotates the idler-wheel which imparts motion to the platter.

As a result, idler-wheel-drive systems not only maintain consistent speed, but they're also rugged, an important asset in many radio production and control-room studios.

The third and newest drive system is the *direct-drive system*, in which the action of the motor (or multiple motors) is transferred directly to the platter (see Figure 5.3). The direct-drive system is by far the most accurate because the speed of the motors is linked to an *oscillator* (a device that produces different frequencies), which is, in turn, controlled by a quartz crystal. The quartz crystal that enables a watch to maintain nearly perfect time exercises the same control over the speed of a turntable. Many of these turntables come with built-in strobe lights and variable speed controls: with them, you can maintain uniform speed or make the platter run faster or slower for special effects.

Most turntables are able to run at two [33-1/3 and 45 rpm (revolutions per minute)] or more speeds. A speed of 16-2/3 is used mostly to play *talking books*, recordings made for the visually impaired. Older machines can also run at 78 rpm, although, outside of children's records, nothing of any consequence is recorded on 78-rpm discs anymore. Even if your turntable can run at this fast speed, you shouldn't use a modern stylus on old 78s. The grooves are huge compared to the microgrooves on today's discs, and a new diamond stylus will roll around in 78 rpm grooves, like a BB in a boxcar. Not only will the sound reproduction be awful, but knicks, bumps, cracks and other imperfections in the disc's surface (remember you're dealing with *old* records) will knock the stylus around terribly, and because the disc's surface is whizzing around under the stylus at such a rapid speed (more than twice as fast as under an LP), damage is inevitable. Unless you have an unlimited equipment budget, you can't afford to ruin styli—which is exactly what you'll do by playing old 78s. These records can be played safely on older turntables, which use steel needles. These needles are big and heavy and were designed for use on 78s. Equipment like this is hard to find outside of a museum though—one of the consequences of advancing technology.

Turntable drive systems are susceptible to three problems:

1. inconsistent platter speed
2. hum caused by a bad ground connection (see your engineer)
3. rumble, when the motor produces a low-frequency vibration that is picked up by the cartridge

If you have a good turntable that's maintained properly, you shouldn't have to worry about any of these annoyances.

5.3 Direct-drive turntable. (Courtesy of Matsushita Electric Corporation of America.)

Disc Recording and Reproduction

It's amazing how many intelligent, otherwise well-informed, people believe that sound is actually trapped in the grooves of a record. In reality, the process of committing sound to disc is similar to, albeit simpler than, tape recording, and although the technology of sound recording made great advances between the the 1920s and 1970s (with the advent of digital sound), the fundamental process remained relatively unchanged.

The Original System

Originally, in Edison's model, a wax cylinder was mounted on a rotating shaft. Positioned above the rolling cylinder was a sharp needle attached to a small diaphragm. The needle was placed on the cylinder and was moved sideways across the cylinder by a screw thread as the cylinder rotated. As sound, focused by a large megaphonelike horn, caused the diaphragm to vibrate, the needle, responding to the vibrating diaphragm, would carve a groove in the wax. Playing the recording meant reversing the process. The needle was placed at the head of the groove and allowed to ride the path it had just cut. The rather violent passage of the groove would cause the needle to vibrate as it had done initially, causing the diaphragm to vibrate as it had done initially, producing a faint sound, which was amplified by the horn: simple, yet brilliant. The cylinder was eventually replaced by a flat disc, which, because it had a greater surface area (not to mention two sides), could hold considerably more information.

The Cartridge

The next great breakthrough was the invention of the *turntable cartridge* (not to be confused with the tape cartridge), which dispensed with the horn, and instead allowed for recording and playing by means of electrical pulses, similar to the process used in tape recording. The cartridge was (and is) a tiny transducer that works on the same principle as the moving-coil microphone. The cartridge houses a coil of wire encircling a magnet. As the stylus rides the groove, its vibrations cause either the magnet or the coil to vibrate (cartridges differ in this respect, although the result is the same). Electrical pulses are thus induced in wires leading from the cartridge to a preamp (preamplifier), which boosts the tiny signal. From the preamp, the signal moves through an amplifier and then to a speaker, which converts the electrical pulses to sound.

The Stylus

The *stylus* is a small conelike chip of very hard material that rides the record's groove and transmits vibration to the cartridge. Styli used to be made of a number of substances, most notably osmium, sapphire, and diamond. Today, the diamond stylus is recognized as the standard because of its durability (it's awfully difficult to wear out a diamond) and its ability to transmit vibration with minimal distortion. The chip is attached to a flexible metal shank, which is mounted on the underside of the cartridge.

The tip of the stylus is cut in one of two shapes: *round* (also called *spherical* or *conical*) or *oval* (also called *elliptical*). Each has strengths and weakness. The spherical stylus doesn't set as deeply into the groove as the elliptical, and, consequently, doesn't come into contact with as much of the groove wall. As a result, the spherical stylus doesn't glean as much information from the groove as the elliptical one. However, because the elliptical stylus sits deeper, it also wears down the record and itself faster—decisions, decisions!

Here's another consideration. If anything tends to take a beating in the studio, it's the stylus, especially the shank. If the shank is bent even a little, the diamond will not sit straight in the groove, resulting in greatly increased record wear and distorted sound. If for no other reason than the cost, stylus durability is very important. A good professional stylus can run over $100, but I've found that the most durable styli aren't the best sounding and the best sounding styli are extremely fragile—more decisions. Forget the spec sheets and propaganda that the manufacturers toss around. When picking a stylus, listen to the advice of experienced producers (they're generally a pretty helpful lot). Above all, listen to what your own ears tell you sounds good.

The Tone Arm

The *tone arm* is a metal bar that supports the cartridge and stylus. It's mounted on a pivot, and, although models vary, it should have these three features:

1. *Tracking control.* The weight of the cartridge and stylus on the disc is very important. If the weight's too great, the disc will wear out faster, and the stylus shank may bend. If the weight's insufficient, the stylus can bounce in and out of the groove and play back distorted sound. The tone arm should have a counterweight on the end opposite the cartridge. Adjust the counterweight so that the arm balances and is perfectly horizontal. Every cartridge has a recommended tracking weight set by the manufacturer. Adjust your tracking

control so that the weight of the cartridge complies with the manufacturer's recommendation. A tracking weight between 1 and 2 grams should suffice. If you have an especially light cartridge, you may have to add some weight to that end of the arm. Taping a paper clip or a penny to the top of the cartridge works pretty well.

2. *Antiskating control.* The record's groove pulls the tone arm toward the center of the disc. Because the groove is so shallow, even a minute force (like when you accidentally bump the turntable) can be enough to cause the arm to jump out of the groove and "skate" inward across the disc. Also, this natural inward pull can cause greater wear to the inner walls of the groove. To counter this force, the antiskating feature puts a slight *outward* pull on the tone arm.

3. *Cuing lever.* This feature prevents the tone arm from being dropped accidentally onto a disc. It's an especially handy option to have, especially if, for example, you want to play a cut that's in the middle of the disc. When engaged, the cuing lever keeps the tone arm a short distance above the platter. You merely move the arm over the proper spot on the record, and when you disengage the lever, the arm slowly lowers onto the disc's surface. Once, during a production session, I overheard somebody say "Cuing levers are for wimps." Mind you, I don't use the cuing lever very often: as long as I can rest my forearm or elbow on something (the turntable housing or the counter on which the turntables are set) while I'm positioning the tone arm, I'm fine. If I can't, I'll use the lever. Let someone else be macho: you take care of your records.

Cuing Records

Being able to start a song or anything else on a disc instantly is crucial to creative production. Fortunately, it's easy to learn.

Let's say you want to cue up a song. First, you need to locate the *precise* spot on the disc where the song begins. Here's the generally accepted procedure (see Figure 5.4):

1. Spin the record, and put the stylus at or near the head of the song.
2. The instant the song begins, stop the turntable.
3. Back up the turntable manually, until you find the start of the song.
4. Finally, continue moving the platter backward an additional one quarter to one half of a rotation.

That's all there is to it. Now, when you start the turntable, the song will begin within one or two seconds.

The last step is important. All turntables need a second or two to get from a dead stop up to 33-$\frac{1}{3}$ or 45 rpm. If you don't allow for a little run-up time—

5.4 Cuing a disc. When the needle hits the start of the cut, stop the disc. Switch the turntable off, and back the platter up a bit to prevent wowing.

that is, if your stylus is positioned right at the start of the cut, your record will *wow* in. The cut will begin before the platter is rolling up to speed, and will begin off pitch, then the cut will slide up to pitch as the turntable reaches full speed. This slide is called *wowing*, and is the sign of a bush-league producer. Different turntables accelerate at different rates of speed (belt-driven models are particularly slow on the pickup), so there's no hard and fast rule about how far to back up the head of the cut. All you can do is experiment with the machines in your studio. Usually, I back the platter ¼ turn for 33-1/3 records and ½ turn for 45s.

Here are some other cuing tips:

1. When you rotate the platter manually, if possible, keep the gear lever (if it's a rim-driven turntable) in neutral. It lessens the pressure on the idler wheel.

2. When you first position the stylus at the start of the cut you want, don't rock the disc back and forth making that "chukka-chukka" sound that rap groups seem to be so fond of. All you're doing is gouging the start of the cut and creating what we call *cue burn* or *cue scratch*. Remember, vinyl isn't very durable to begin with, and when you drag a diamond chip across it, the chance for damage

is great. The stylus does not point straight down. It actually angles back very slightly toward the cartridge, so when you rock a disc back and forth, especially if it's the cheaper grade of disc that record companies generally send as freebies to radio stations, one or two passes and the stylus will eat into the record like a rat through cheese.

3. Along the same lines, some turntables have a small recess in the platter, and a 45-rpm record will sit in this recess, slightly below the surface of the platter. Because of this shallow well, the angle at which the stylus contacts the surface of the 45 is even greater, thus increasing the chance of damage. One solution is to put a junk 45 down in the well first and then put your good disc on top of it. Your good disc will now be at the surface level of the platter. The only problem with this is possible slippage between the discs, so when cuing, be prepared to back the platter up more than the usual half turn.

4. Some people like to keep the turntable locked into gear when they cue a record and then manually move the disc instead of the turntable. By applying pressure near the center of the record, this technique works well, especially with 45s—as long as you never let your fingers touch the record's grooves. Also, a felt pad on the platter will allow the disc to slip and to be cued more easily. Many turntables come with a pad already affixed; if your machine didn't, you have a couple of options: (1) you can pay $20 or more and buy one from a professional supply house, or (2) you can pay less than $3 and buy some medium weight felt from your local fabric store and cut a pad to fit your platter (Don't forget the spindle hole!).

Warning!! A lot of discount department stores sell plastic foam turntable pads. Don't bother. Not only do they not slip very well, when you slipcue (see the next paragraph), they bind up under the discs and can scrape them. The people who make these pads say that felt turntable pads generate static electricity and that the foam pads don't. It's true, cuing with felt pads can cause static, especially in winter months, but I've found the problem to be minimal— just like the value of the foam pads.

5. Slip-cuing is one of the first tricks a budding disc jockey learns (see Figure 5.5). When you find the beginning of your cut, only rotate the platter back about an inch. Then start the turntable, but hold the disc in place by pressing against the disc's edge. When you release the disc, since the platter has been turning and is already up to speed, the record almost instantly jumps up to speed. For split-second timing, the slip-cue is hard to beat, and if you want to begin in the middle of a cut, the slip-cue can minimize (and sometimes eliminate) any wowing.

As easy and useful as the slip-cue is, it unfortunately ties up one of your

5.5 Slip-cuing. Lightly hold the disc in place while the platter turns. Release the disc with a lateral movement.

hands. Set the stylus at the proper position, leave it there with the turntable switched off, and go about other business until you're ready to slip-cue. For flawless execution, remember two things: First, maintain just enough pressure on the disc to keep it from turning with the platter; too much pressure can slow down the turntable. Second, when you release the disc, move your hand outward laterally from the disc: if you release the disc with an upward or downward motion, you risk jarring the record, and bumping the stylus out of the groove. As you might guess, the platter *must* be able to slip easily beneath the record, so a felt pad is a necessity. Of course, keep your fingers away from the grooves.

Turntables and Records: Care and Handling

Bear in mind the following tips for the care and handling of your turntables and records:

1. Keep the stylus free of dirt. *Do not ever* scrape the stylus with your finger. Novices in my classes sometimes scrape the stylus with a fingertip to see if they're getting a signal through the speakers. The only thing they get is a loud scream from me. The best thing to use to clean the stylus is a short-bristled nylon brush. If you don't have one and can't buy one, you can usually

get most of the dirt off of the stylus by blowing it off. One or two short, sharp puffs should do the trick. If you're not sure your turntable is on, set the stylus on a disc—not your finger!

2. Periodically check the stylus shank. If you see that it's bent and the diamond looks askew, get rid of the stylus. If it's bent, and you've only had it a short while, make sure your antiskating and tracking adjustments are properly set.

3. Never touch the grooves of a record with your bare hands—*ever*! That includes during removal from and replacement in the record's jacket. No matter how clean your fingers appear to be, there's always oil on them. One of my students, as a challenge, scrubbed his hands almost raw, and dried them under a blow-dryer. His hands were clean enough for brain surgery, but when he handled a record improperly, his fingerprints were plainly visible on the vinyl.

4. Never stack good records on a turntable.

5. Never keep records near a heat source, due to the possibility that they will warp. Once a disc is warped, consider it gone. It's true that if the warp isn't too bad, and if it's mostly away from the edge of the disc, you can sometimes flatten the warp out using weights. However, unless it's one of your favorite records and is irreplaceable, it's not worth the effort.

Compact Discs

The latest advance in turntable technology involves the compact disc (see Figure 5.6). A discussion of digital sound is in order. Since the beginning of recorded sound, producers have used *analog* methods of recording. This means that the original sound is converted to an analogous form, such as the electrical pulses in a mike cord, or the groove on a record, or the magnetic patterns on a piece of tape. Looked at a different way, analog recording is like translating from one language to another, or like the *Cryptoquote* feature in the newspaper, in which a famous quotation is put into a coded form for the entertainment of those who can figure it out and for the annoyance of those of us who can't.

Linguists will tell you that no language translates perfectly into another. Each has its own quirks and cultural idiosyncracies. This is especially true of poetry, where shades of meaning and actual sound (like rhyme) may not shift well from one tongue to another. Analog recording also has its flaws. When you transduce a sound into another form, the shift is never 100% perfect. Distortion is one pitfall. Another is noise from the equipment you're using to provide the transduction (electrical hum, tape hiss, etc.). Modern recording techniques

5.6 A professional CD player. (Courtesy of Numark Electronics Corp.)

and improved equipment have done much to whittle these problems down to the point where there not as noticeable as they used to be, but they're still there.

Enter digital sound—in the late 1970s, an entirely different process was developed in which the sound, instead of being converted into an analogous form, which could be flawed, was converted into a purely numerical (digital) code (which could also be considered an analogous form, but which is numerically measurable and not subject to the continuous fluctuations of analog forms). The result was virtual perfection. Here's how it works.

Imagine 1 second of recording tape (7-½ inches) laid out horizontally. Now imagine cutting that piece of tape into 50,000 tiny vertical slivers. Each one would be only 0.00015-inches wide. The next step would be to examine each sliver and assign it a numerical (digital) code based on what's imprinted in it. Each tiny piece of tape is evaluated in this fashion, with the numerical code changing to reflect even the slightest variation of signal change. This, in essence, is how sound is converted to a digital format. The signal is first *sampled*— that is, broken into tens of thousands of pieces *every* second. The rule of thumb used to be that the number of samples per second should be no less than double the highest frequency contained within the sound. Thus, if you're recording music that has harmonics around 25 kHz, you'd want a sampling rate of at least 50,000 samples per second.

Once sampled, each bit of sound is assigned a digital code using binary numbers. *Binary math* means that two digits (0 and 1) are used in various combinations to represent *any* number, regardless of length. In regular arithmetic, the place values of each number are (from right to left) one (units), ten, one hundred, one thousand, ten thousand, etc. In binary math, the place values are one (unit), two, four, eight, sixteen, etc.; each place is double the one previous. In each place, a 1 or a 0 indicates whether or not that place's value is part of the number you have in mind. For example, the number 5 is represented as 1-0-1, that is, one four, no twos, and one one. 14 is 1-1-1-0; one eight, one four, one two, and no ones. The number of places is infinite—just keep doubling the value of the previous column. In binary, 325 is written as 1-0-1-0-0-0-1-0-1:

256	128	64	32	16	8	4	2	1
1	0	1	0	0	0	1	0	1

$$256 + 64 + 4 + 1 = 325$$

So the binomial (or binary or digital) equivalent of 325 would be 1-0-1-0-0-0-1-0-1. Using this system, each bit of sound in a signal is assigned a code number, the more complex the sound, the longer (more complex) the code. Each code for each fraction of each second of sound is stored electronically in the equipment or on special digital audio tape (DAT). When the information is played back, specialized equipment reads the numbers and re-creates the sound equivalents without any of the glitches that analog recording is heir to.

The CD uses this technology. Instead of a groove (analogous to the original sound), microscopic pits are carved into the disc's surface and traced by a laser. As the laser passes over the pits, a sequence of digital information (1s and 0s, basically) is fed into the disc player and converted into sound.

The disc itself is 4.7 inches in diameter, and because the pits are microscopic, more information can be compressed on one side of a CD than onto both sides of a standard LP. Although CDs are currently recorded on only one side, it's just a matter of time before the two-sided CD is a reality.

Another difference is the sequence in which the information is gleaned from the disc. With an LP, the tone arm is set on the outside edge of the disc, and progresses inward as the disc revolves at a fixed rate of speed. With the CD, the laser begins reading at the innermost part of the disc and moves outward. Also, the speed changes. When the laser is at the beginning of the disc (the center), the disc revolves at 200 rpm. As the laser moves outward, the speed of the disc gradually increases until, when the laser has reached the outermost portion of the disc, the speed has increased to 500 rpm.

It's no wonder more and more studios are switching to CDs. The sound is magnificent: wide frequency response, no perceptible distortion, virtually perfect fidelity, and you never have to replace expensive styli. Also, you don't have to worry as much about getting oil from your hands on the grooves (and the stylus). What's more, because of its special coating, the CD won't warp, scratch, skip, or wear out with normal use, although, because the technology's so new, its performance and durability over time have yet to be determined. Indications are that some degradation of sound quality does seem to occur, but only the passage of time will yield definitive answers.

Because of the CD, the vinyl disc seems to be headed into oblivion. A number of record companies have already cut back or eliminated their vinyl disc production in favor of the CD. More and more music and sound-effect libraries are coming out on CD. As a producer, though, one factor keeps albums in my studio—the fact that a CD can't be handled and manipulated as can an LP. True, you don't have to cue a CD: you merely press a button. However, if you want to lengthen the attack time, you're stuck. With an album, you can just back it up a revolution or two. Also, if you want to take an excerpt from the middle of a cut, you'll have a tough time doing so with the average CD player. CDs are great, but I don't think you should surrender total control to any machine, so don't throw your albums out just yet.

Review

1. A turntable consists of a *drive system*, which moves the platter, and an *electronic system*, which reproduces sound from the movement of the stylus through the groove.
2. The three drive systems available are belt drive, idler-wheel (rim) drive, and direct drive. Belt-driven turntables are rarely used today.
3. Most professional turntables run at speeds of 33-1/3 and 45 rpm (revolutions per minute). An older machines may have a setting for 78 rpm. Special turntables for playing talking books can run at 16-2/3 rpm.
4. Turntables are susceptible to three problems: inconsistent platter speed, electrical hum, and low-frequency rumble. Direct-drive turntables eliminate most of these problems.
5. A record's groove is cut by a sharp tool vibrating in sympathy with a sound. When a needle rides the same groove and vibrates in an identical fashion, the original vibration is re-created.
6. The turntable cartridge converts the needle (stylus) vibration into electrical pulses, which are sent to an amplifier and then to a speaker, which converts the pulses back into audio vibration.
7. Today's styli are generally chips of diamond and can be either conical or elliptical in shape.

8. Tone-arm movement is controlled not only by the record groove, but also by tracking and skating controls.

9. Cuing records is necessary to start a particular cut at a precise moment. Improper cuing technique can result in damage to records.

10. Slip-cuing allows for a cut to start instantaneously.

11. Keep discs and styli free of dirt and oil.

12. CDs utilize digital (as opposed to analog) recording technology and produce sound that has almost perfect fidelity. A sound is sampled thousands of times per second, converted to binary numbers, and stored. When retrieved, the numbers are converted back into the original sound.

13. The CD is imprinted with binary information, which, in turn, is read by a laser.

14. CD durability is still in question.

6

Loudspeakers

You only need to know two things about your speakers: how they work and how to position them. I can explain the former but can only give you guidelines for the latter.

Speaker Construction and Operation

The most common type of studio speaker is the moving-coil model. As the name suggests, it works much like a moving-coil mike—only in reverse. If you could remove the foam from the front of the speaker cabinet, you'd see two or more individual cones, each of which is technically a speaker. A fixed natural magnet, shaped something like a horseshoe, is mounted inside each speaker, and within the fixed magnet, a floating electromagnet is mounted. An *electromagnet* is one which is man-made by wrapping an electrical current around a metal core, as opposed to a *natural magnet*, which exists without any help from us. In studio speakers, the electromagnet is activated by the current coming from the console amplifier. As the two magnets interact, the electromagnet, which is not fixed, vibrates in response to the varying electrical current. This vibration is transferred to a paper cone, which in turn transmits the vibration to the air in the form of the sound waves we eventually hear.

Let's follow the process from moving-coil mike through to the moving-coil speaker:

1. *Sound* (acoustical energy) makes the mike diaphragm vibrate (change to mechanical energy).
2. The moving diaphragm causes the attached magnet to translate the vibration into electrical pulses (change to electrical energy), which are sent to the preamp (recall that mikes produce low-level signals that need this boost) and then to the amplifier.
3. From the amplifier, the current is sent to the speaker's electromagnet (change back to electromagnetic energy), which then vibrates in response (change back to mechanical energy).

4. The vibration is transferred to the paper cone, which imparts the vibration to the air as sound waves (change back to acoustical energy).

We have a chain of transducers, which changes the original sound vibration into a number of corresponding energy forms, culminating in the recreation of the original vibration.

If we remove the front of a speaker cabinet, we generally see an array of at least two cone-shaped horns. The larger horn, usually positioned near the bottom of the cabinet, is called a *woofer*, and each smaller horn is called a *tweeter*. The need for multiple horns is evident when you realize that no single horn can accurately reproduce the vibrations over the entire sound spectrum we're able to perceive. If you've ever heard an old gramophone recording played through one of those huge metal *morning glory* horns, or even if you've heard someone speak through a megaphone, you know that the original sound isn't like what you hear. The rule is, the larger the horn or cone, the better able it is to reproduce lower frequencies; however, the larger the horn or cone, the *less* able it is to reproduce higher frequencies. So instead of running complex vibrations through a single horn, today's speakers come equipped with a *crossover network*, a system of filters that takes an incoming signal and splits it at a preset *crossover frequency*. Everything above the crossover frequency is shunted to the tweeter, and everything below it goes to the woofer. What we have here is known as a two-way speaker system (see Figure 6.1), with the crossover frequency generally set somewhere between 500 and 1000 Hz. As good as this system is, the funneling of everything over 500 Hz (extending well beyond 20,000 Hz) through a single tweeter presents some of the same problems that the two-way system was designed to prevent, albeit on a smaller scale. So

6.1 An array of two- and three-way speakers. (Photo courtesy of UREI/JBL Professional.)

6.2 Some of the best speakers in the world are in headphones. But don't use them as your monitor system unless you're sure your audience will do the same. (Photo courtesy of Sennheiser Electronic Corporation.)

manufacturers introduced a three-way system consisting of a woofer and *two* tweeters, one for lower treble frequencies (500 Hz to 5 kHz) and the other for the upper treble range (above 5 kHz). If you have the money to throw around, four-way speakers are available, as well as other systems featuring dizzying arrays of horns and internal speaker controls. For the money *and the sound*, however, a good three-way speaker (see Figure 6.1) should fill all your monitor needs.

Monitor Placement

Your monitors are the means by which you can hear what you're doing: what's going onto and coming off of your tape and whether your final mix is good. In case you think that headphones, or *phones*, are a good substitute, you're sorely mistaken.

I love headphones (see Figure 6.2). Listening to a fine recording through a high-quality set can really make you feel like you're sitting in on the session, *but*, unless the audience for your production is going to be using phones, don't use them as your monitor system. With phones, there's none of the interaction between the sound and the environment that shapes our perception of the sound. The phones put any stereo images inside rather than around your head. As a result, a sound mix heard through headphones will not sound the same when heard through loudspeakers. It's obviously of paramount importance to know how your product is going to sound to your audience, so unless your studio has poor acoustics with lots of reverb, use phones only when you have to, such as when you're recording out on the street, or when your monitor amp blows a fuse.

As a rule, you should position your speakers so that you hear all frequencies reproduced clearly (see Figure 6.3).

6.3 Speakers should be placed at or near ear level, and angled toward the operator.

1. If you can position your speakers at ear level, do so. If you have to mount them above ear level, such as on a wall, try to angle them down toward you. High frequencies are very directional. While playing some music, turn your speakers 90° and hear what happens to the treble.

2. Position the speakers at least 3–4 feet away from you, and 4–6 feet apart. If your speakers are too close or too distant, the depth of stereo images can be distorted or, in extreme cases, lessened. Also, higher frequencies are hard to hear over distance because of their inherent lack of power.

3. If you stand your speakers on end, with the tweeter on top, remember again that high-frequency sound waves are comparatively feeble and can, especially if the speakers aren't angled toward you, float right over your head. To ensure that these waves reach you relatively intact, lay your speakers on their sides, with the woofers toward the center of the console, and the tweeters toward the outside. This gives your highs a fighting chance to reach you without being overwhelmed by the more powerful bass tones.

Above all, remember that speaker placement is very subjective, depending on the acoustical makeup of your studio and the predisposition of your own ears.

Review

1. The typical speaker works like a moving-coil mike, but in reverse, transducing an electrical signal into mechanical movement of the speaker's cones, and then transducing this movement into sound.
2. A speaker contains woofers and tweeters. *Woofers* reproduce low frequencies, and tweeters reproduce midrange and treble frequencies.
3. The crossover network separates the signal into frequency bands and then sends the bands to the individual horns (woofers and tweeters), which reproduce them as sound. Multiple horns are needed, because no single horn can reproduce all frequencies.
4. The point at which a signal is divided is called a *crossover frequency.*
5. A good three-way speaker (one woofer, one midrange tweeter, and one treble tweeter) is sufficient for studio use.
5. Speakers should be positioned at least 3–4 feet away from you and 4–6 feet apart. If possible, position them at ear level. If this isn't possible, at least angle them down (or up) toward you.

7

□ □ □
□ □ □
□ □ □

The Patch Panel

For some reason as yet unknown to me, the sight of a patch panel has the same effect on many novice producers that sunlight has on vampires. Nonetheless, understanding and mastering the patch panel (sometimes called a *patchbay*) marks our first real move into creative production because it allows us to manually control and alter the flow of a signal to suit our needs.

Imagine this. You have a microphone wired to a channel input on your console. What happens if that input malfunctions, or if the key switch won't work, or if someone accidentally spills spaghetti sauce on the pot and short-circuits the channel? (I've seen it happen!) In any and all of these cases, the microphone, because it's tied directly to the channel (hard-wired), could not be used.

Hard-wiring would also prevent more than one signal from reaching a single channel on a piece of equipment. You could, for example, record music from a disc onto a cassette, but you could not talk over the music at the same time because only one of the signal sources, in this case either the mike or the turntable, could be wired directly to the cassette recorder.

Overview: How Patch Panels Work

The patch panel (Figure 7.1) is inserted between the studio's equipment and the console, so that signals leaving the console meet the patch panel before arriving at any of the equipment; similarly, signals leaving any of the equipment hit the patch panel before reaching the console. Regardless of the signal's direction, whenever it reaches a patch panel, you can intercept it and change its path.

The key to understanding and using the patch panel is a familiar one: *Follow the signal's path. Where is the signal coming from? Where do you want it to go?* In most patch panels, the upper row of sockets is where the signal is coming from, and the lower row shows where the signal is going to (Figure 7.2). The sockets are arranged in vertical pairs and are connected internally, so that current flows from the upper socket to the lower. Thus, a signal enters an upper socket

7.1 Standard 48-socket patch panel. (Courtesy of PRO-CO Sound, Inc.)

and is channeled into the lower one (see Figure 7.3). For example, even though we may have an input on the console labeled *cassette input*, the output of the cassette player is *not* wired directly to that console input. If you look behind your gear and actually follow the wiring, you'll see that, in this example, a wire runs from the cassette output to an upper socket on the patch panel. Then another wire runs from the lower socket to the cassette input on the console. So the cassette player is indeed connected to the console, but not directly.

The path that the current naturally follows is called the *normal path* or, simply, the *normal*. The cassette player is said to be *normaled* to the upper socket of the patch panel: the upper socket is normaled to the lower socket, and the lower socket is normaled to the console. A normal can be a single length of wire or a series of things. For example, if you plug a lamp into a socket, the normal path the current follows is from the wall to the lamp. Now, let's plug the lamp into an extension cord and the cord into the wall. Even with the addition of another component (the extension cord), the current still terminates at the lamp. The normal hasn't changed. It doesn't matter how long the path is or how many pieces or segments it comprises—what's important is that the flow be uninterrupted.

	Prog		Prog		Prog		Prog		Prog		Prog		Mult		Mult		Mult	
Out	○	○	○	○	○	○	○	○	○	○	○	○	○	○	○	○	○	○
	L	R	L	R	L	R	L	R	L	R								
In	○	○	○	○	○	○	○	○	○	○	○	○	○	○	○	○	○	○
	T1/1	T1/3	T1/2	T1/4	T2		Cart		Cas 1		Cas 2							

	T1/1		Cart		T1/2		Cas 1		T1/3		T2		T1/4		Cas 2		Eq	
Out	○		○	○	○		○	○		○	○	○		○	○	○	○	○
	L	R	L	R	L	R	L	R	L	R	L	R	L	R	L	R	L	R
In	○	○	○	○	○	○	○	○	○	○	○	○	○	○	○	○	○	○
	Ch 5/1		Ch 5/2		Ch 6/1		Ch 6/2		Ch 7/1		Ch 7/2		Ch 8/1		Ch 8/2			

7.2 Patch panel array. The upper panel carries outbound (from the board to the gear) signals, and the lower carries inbound (from the gear to the board) signals.

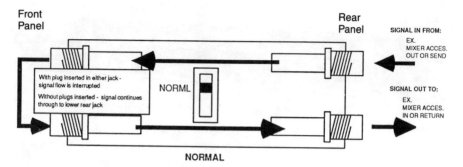

Front Panel

Rear Panel

SIGNAL IN FROM:
EX.
MIXER ACCES.
OUT OR SEND

With plug inserted in either jack - signal flow is interrupted

Without plugs inserted - signal continues through to lower rear jack

NORML

SIGNAL OUT TO:
EX.
MIXER ACCES.
IN OR RETURN

NORMAL

7.3 How a patch panel works. A signal enters through the rear of the panel at an upper socket, moves to the upper socket at the front, is transferred to the front socket directly below, and exits the unit through the lower rear socket. (Courtesy of PRO-CO Sound, Inc.)

Your patch panel should come with a number of patch cords. These cords will have either one (Figure 7.4) or two plugs (Figure 7.5) on each end. The double cords are just like two of the single cords tied together: a single cord channels a single signal, and a double cord channels two signals, frequently the left and right sides of a stereo signal. The only problem with the double cords is keeping track of which plug on one end is connected with which plug on the other end. To simplify things, each of the double plugs has a single serrated edge on its plastic housing. The plugs located by the serrated edges are attached and are like the opposite ends of a single cord. To completely avoid the confusion, stick with single-plug patch cords. They're easier to use and afford you greater flexibility.

When a patch cord (or more simply, a *patch*) is inserted into a patch panel

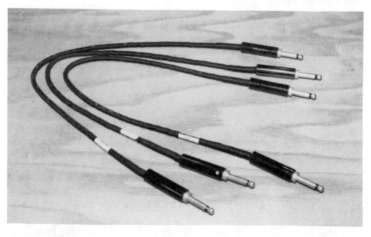

7.4 Single-plug patch cords. (Reprinted with permission of Switchcraft, Inc., a Raytheon Company.)

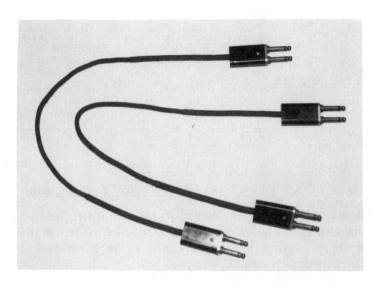

7.5 Double-plug patch cords. (Reprinted with permission of Switchcraft, Inc., a Raytheon Company.)

socket, the normaling between the two sockets is broken. *A patch always overrides the normal.* In the previous example, a patch into the cassette output socket on the panel would break the normal, and the signal, instead of flowing to the lower socket and then to the console input, would instead flow out into the cord. The free end of the cord could be plugged into a different console input on the panel, and the signal would therefore move into the console through an entirely different input. This is how you can get around a broken pot: patch the output of the piece of equipment into a different console input (see Figure 7.6). When you do this, you create a new normal, which will last as long as the

	Prog		Prog		Prog		Prog		Prog		Prog		Mult		Mult		Mult	
Out	O	O	O	O	O	O	O	O	O	O	O	O	O	O	O	O	O	O
	L	R	L	R	L	R	L	R	L	R								
In	O	O	O	O	O	O	O	O	O	O	O	O	O	O	O	O	O	O
	T1/1	T1/3	T1/2	T1/4	T2		Cart		Cas 1		Cas 2							

	T1/1		Cart		T1/2		Cas 1		T1/3		T2		T1/4		Cas 2		Eq	
Out	◉		O	O	O		O	O		O	O	O		O	O	O	O	O	O
	L	R	L	R	L	R	L	R	L	R	L	R	L	R	L	R	L	R
In	O	O	O	O	◉	O	O	O	O	O	O	O	O	O·O	O	O	O	O
	Ch 5/1		Ch 5/2		Ch 6/1		Ch 6/2		Ch 7/1		Ch 7/2		Ch 8/1		Ch 8/2			

7.6 Patching to change inputs. Tape 1, track 1, which is normaled to a channel 5 input, is patched to a channel 6 input.

7.7 Making a mono recording of a signal that's only on one side. The signal on the board's left program output line is normaled to the cassette's left side. To get the signal also to the cassette's right side, patch from any Prog L Out except the one feeding the cassette's left input (X).

patch remains in place. Remember that the patch overrides everything it comes in contact with. If you patch the cassette output into the tape recorder input on the console, the signal from the cassette will not only come up on the tape recorder input, but will also block any signal from the tape recorder from reaching the input. If you want to hear the tape recorder, you'll have to patch it into a different input on the console.

Patching not only affects signals going into the console (inbound), but also those coming from the console (outbound) to the pieces of equipment. Except for mikes and turntables, much of your gear is made to send *and* receive signals. The patch panel is meant to provide a bridge for signals traveling in both directions. A check of your panel should reveal a number of upper row sockets labeled *program out* or *console out*. Below them are sockets with the names of your various pieces of gear or tape tracks. The signal from the console is sent (frequently by a device called a *distribution amplifier*) to each of the program- or console-out sockets and then to the equipment. In other words, the signal coming from the console is duplicated at each of the console outputs on the patch panel before being sent on to the specific track or machine. With a system like this in place, a number of signals can enter the console from different sources, be combined into a single signal, and sent out anywhere the producer desires (see Figure 7.7).

The patch panel also sends signals to the left and right. Although I don't want to get into stereo just yet, it's important to note that normaling also affects whether a particular machine or track will play on the left or right side of your system. If your studio is wired for stereo, the normal path will also include a left or right designation (or both if it's a mono signal) for every piece of gear or track in your studio. In tape equipment with only two tracks, it's likely that one track will go left and the other right. A mike may be normaled to a certain side (see Figure 7.8). In a multitrack tape deck, some tracks may be wired left, and others right. It all depends on your needs—and the imagination of your engineer.

Multiples

Your patch panel should also have a number of *multiples* (see Figure 7.9). A multiple is a group of sockets (generally 4) wired to each other and not to any equipment. A signal patched into any one of the sockets is instantly duplicated at the other three. A multiple (generally called a *mult*) is handy, for example, if you have a signal on either the left or right side and want it to play out of both sides for a mono recording. The signal is patched from its source *into* any of the mult sockets (None of the mult sockets is labeled *input* or *output*. The sockets are equal. Whichever socket the signal is patched into becomes the input, and the other three sockets become outputs.) The signal is automatically duplicated at each of the three remaining sockets. Patches are inserted into any two of the sockets, and with each carrying a copy of the original signal, they're patched *from* the mult *into* two console or equipment inputs, one patched to the left and the other to the right.

How to Use Multiples

A mult can accept only one signal. (Mults are easy to wire. Ask your engineer to rig three or four mults for you, and then take her or him to lunch.) If you patch two signals into a mult, the mix coming out will be distorted.

	Prog		Prog		Prog		Prog		Prog		Prog		Mult		Mult		Mult	
Out	○	○	○	○	○	○	○	○	○	○	○	○	○	○	○	○	○	○
	L	R	L	R	L	R	L	R	L	R								
In	○	○	○	○	○	○	○	○	○	○	○	○	○	○	○	○	○	○
	T1/1	T1/3		T1/2	T1/4		T2		Cart		Cas 1		Cas 2					

	T1/1		Cart		T1/2		Cas 1		T1/3		T2		T1/4		Cas 2		Eq	
Out	●		○	○	○		○	○		○	○	○		○	○	○	○	○
	L	R	L	R	L	R	L	R	L	R	L	R	L	R	L	R	L	R
In	○	●	○	○	○	○	○	○	○	○	○	○	○	○	○	○	○	○
	Ch 5/1		Ch 5/2		Ch 6/1		Ch 6/2		Ch 7/1		Ch 7/2		Ch 8/1		Ch 8/2			

7.8 Patching to change sides. This signal is normaled to the left side of the channel 5 input, but via a patch, is fed to the right side of the same input.

7.9 A multiple (mult) is used to make copies of an input signal.

So if you have two or more inbound signals that you want to mix and split via a mult into a mono recording, there are two ways to do it:

1. Split the *individual signals* on their way to the console (see Figure 7.10). If you have a number of mults, each signal can be sent into a separate mult, then split, and then sent to inputs on the console.

2. Split the *mix of the signals* as it comes out of the board (see Figure 7.11). Bring all the signals into the console on the same side. Then, using one of your *program-out* or *console-out* patches (they're all the same!) patch *the output of the board* (all the signals combined) into one mult, split the signal, and patch it into a stereo or a 2-track tape machine (reel-to-reel, cart, cassette).

This second option uses only one mult and therefore fewer patch cords. Bear in mind two things: the inbound signals *must* be on the same side, left or right (if you've got two left-side signals and one right-side, use a patch to transfer this last signal to the left side), and make sure you don't try to patch *out* of the console, into a mult, and then back into the console—the system will feed back. Remember that the signal path out of the console is *console to patch panel to tape*. Whether a signal is outbound or inbound, you can channel

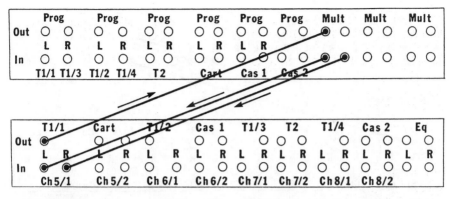

7.10 Splitting an inbound signal with a mult. Here, tape 1, track 1 is normaled to the left side of the channel 5 input. Using the mult, the signal is split and sent into both sides of the input.

it into a mult, but you can't make a u-turn in the patch panel, and send the signal back where it came from.

Can you patch from one piece of equipment directly into another? Yes (see Figure 7.12)! If you want to record a reel directly onto a cassette, you can patch from the reel-to-reel recorder's output to the cassette recorder's input, and the signal will travel from one machine to the other. However, because the patch has made the path perfectly direct, the console has been bypassed—the signal isn't reaching the board. You won't hear anything unless you plug your headphones into one of the machines you're using.

7.11 Splitting an outbound signal with a mult. Any program left output will do because they all contain the same signal. Actually, sending this signal to both sides of the cassette can be accomplished without using a mult (see Figure 7.7).

	Prog		Prog		Prog		Prog		Prog		Prog		Mult		Mult		Mult	
Out	O	O	O	O	O	O	O	O	O	O	O	O	◉	◉	O	O	O	O
	L	R	L	R	L	R	L	R	L	R								
In	O	O	O	O	O	O	O	O	◉	◉	O	O	◉	O	O	O	O	O
	T1/1	T1/3	T1/2	T1/4	T2		Cart		Cas 1		Cas 2							

	T1/1		Cart		T1/2		Cas 1		T1/3		T2		T1/4		Cas 2		Eq	
Out	◉		O	O	O		O	O		O	O	O		O	O	O		O
	L	R	L	R	L	R	L	R	L	R	L	R	L	R	L	R	L	R
In	O	O	O	O	O	O	O	O	O	O	O	O	O	O	O	O	O	O
	Ch 5/1		Ch 5/2		Ch 6/1		Ch 6/2		Ch 7/1		Ch 7/2		Ch 8/1		Ch 8/2			

7.12 Using the patch panels to bypass the board. The signal from tape 1, track 1 is sent via a patch into a mult, split, and then sent directly into the left and right of the cassette. Because the signal never reaches the board, if you want to hear what you're doing, you'll have to plug your headphones into the cassette deck.

Before you begin work in the studio, it's always a good idea to check the patch panel and to remove any patches you're not going to use. It's not unusual for a beginner to reach new heights of frustration trying to bring a signal up through the board, only to have the effort thwarted by an overlooked patch cord that's blocking the signal's path. Whatever you do, pull the patch out by the plugs—don't yank the cord. It's not indestructible, and an intermittent patch cord is about as useful as a square wheel. Finally, if your patch connections don't seem really solid, that is, if the connections are intermittent or make *snap-crackle-pop* noises, take some steel wool to your plugs before you have your engineer rip the patch panel apart. Very often they are made of brass, and the air can cause a coating of oxide to form on them. This oxide, since it's not a good conductor of electricity, will play hob with your connections. Plain steel wool (not the kind with soap in it) should do the trick.

There's nothing magical, mysterious, or mortifying about a patch panel. If the sockets are clearly and properly labeled, all you have to do is keep track of the signal's path. The best way to become acclimated to your patch panel, and to your system in general, is to play with it. Patch signals from machine to machine, from track to track, from side to side. One of the nice things about production is that the more time you spend playing in the studio (and it really is play), the faster and more thoroughly you're going to learn. If only the rest of life could be like that.

By now, you should have a good grasp of how your studio's equipment works. If this were going to be purely a technically oriented production text, we'd be finished, but it's not, and we've really just begun.

Review

1. The patch panel serves as a bridge between the studio's equipment and the console. Signals traveling to or from the console pass through the patch panel.
2. The patch panel's sockets are arranged in vertical pairs. A signal enters through the upper socket and exits via the lower socket. Usually, the upper socket determines where the signal comes from, and the lower tells where the signal's headed.
3. A *normal* is a path through which a signal naturally flows. A normal can consist of a single length of wire, or a number of segments—as long as the flow is unimpeded.
4. Plugging a patch cord into a socket breaks an existing normal and establishes a new one. This allows you to connect inputs and outputs that aren't actually wired together.
5. A patch always overrides a normal.
6. A patch can be used to send a signal to the left or right for creating stereo images.
7. A multiple (mult) is a set of four interconnected sockets. A signal input to one socket will output from the other three, creating three copies of the original signal. A mult can accept only one incoming signal.
8. If necessary, the output of one machine can be patched directly to the input of another, thus bypassing the console.

8

□ □ □
□ □ □
□ □ □

Using Your Voice

Do you remember the first time your heard your own voice played back from a tape? If you're like most people, your first reaction was shock and your second, depression: shock, because the voice you thought you've known all life long, you haven't really known at all; and depression, because even though you might not have the greatest voice in the world, could it really be as lifeless as that droning coming from the speaker? Yes, indeed, the voice from the speaker is what you really sound like. Your skull is honeycombed with nooks and crannies, and when you speak, the sound reverberates and resonates throughout this cranial space, coloring the tone significantly. That's what *you* hear. Unfortunately, everyone else hears the sound coming directly from your mouth, without any added coloration. That's your true voice, and contrary to what you might feel, it's probably not as bad as you think; you were just under the impression it was a lot better.

Now that you've come to grips with the reality of your voice, let me make two important points:

1. If you weren't blessed with a voice like Orson Welles, or Mason Addams, or Glynis Johns, or Tammy Grimes, or James Earl Jones, there's nothing you can do about it. If you have a reedy tenor voice, all the practice in the world isn't going to make you a booming bass. The voice you were born with is the only one you've got, no more, no less; great voice talents, like great sprinters, are born, not made. Yes, you can help a runner improve his or her start, stride, and strategy, but if the raw ability isn't there to begin with, the runner will never be a world champion.

2. You *don't* need a *great voice* to be an *effective communicator*! What's important is learning to use correctly the voice you have! Your voice can be trained to sell products or ideas, and to motivate people, simply by mastering and practicing a few skills, and by realizing that your voice is a unique, living tool, not merely something you've been carrying around in your throat all your life.

Occasionally, in some radio stations, the production manager is strictly a technician, and the voice work is provided by the on-air staff or outside talent. In most situations, the production head has to record a lot of voice tracks. If you're in radio production, you'd better have a decent set of pipes. If your voice isn't up to snuff, this chapter should be committed to memory.

Four Variables of Voice Technique

Learning to use your voice properly is not easy. It requires a lot of time, effort, and concentration, but it's a fairly simple process because there are only four variables involved.

1. your pitch
2. your loudness
3. your pace
4. your emphasis

Pitch

I've never heard anybody who's a true *monotone*—that is, whose voice consistently remains on one pitch—but we all know someone (frequently a teacher) whose voice is so boring it puts us, and everyone else within earshot, to sleep. This person, although not truly monotonic, uses only a fraction of his or her pitch range, but, to be fair, few of us really use the pitch range we have available. Get to a piano and find the highest and lowest notes you can sing. Most people will have a range around 1-½ octaves (11 or 12 consecutive white keys). A decent singer can cover two octaves (15 keys) or more. Yet, the frequency range we use during everyday speech centers around 4 or 5 keys. In other words, we only use a small portion of the pitch range we have available to us.

Learn to stretch your frequency range.

Be aware that your voice naturally moves up and down in pitch as you speak. When your voice pitch moves up, make it go a little higher. As your voice moves down, nudge it a little lower. The key word here is *little*. Your aim is to slightly stretch out your frequency range—don't force anything. Adding the equivalent of only a note or two to both ends of your range can increase your overall speaking range 50–100%. If you push too hard and overstretch, you'll sound ridiculous. Remember newsman Ted Baxter on the old "Mary Tyler Moore Show," who, when he was on the air, tried to shove his voice down into his socks? Or what about those screaming 1960s rock jocks, whose voices meandered all over the place? Were they funny? You bet. Entertaining? Sure. Would you use them to sell your product or idea? I don't think so. Where would you rank the credibility of voices like those? In special situations, a crazy voice may be called for, but that's the exception, not the rule.

Loudness

As with pitch, we use only a fraction of the dynamic (loudness) range we were born with. In fact, a boring speaker is more likely to be *mono-dynamic* (one loudness) than *monotonic* (one pitch). The pitch range we employ in everyday speech stays relatively constant, but our loudness obviously varies with the situation—and it certainly can vary! With casual conversation, we use a decibel level near the bottom of our range. Compare that level with the level you'd use addressing a small class, or addressing a large class, or speaking to a crowd in an auditorium, or trying to get the attention of someone on the other side of a room full of noisy New Year's Eve celebrants at one second past midnight.

Unless otherwise dictated, the level you use on-mike should approximate the loudness you'd use talking casually with *one* person. Relax, and don't yell. It's incredible how a novice's decibel level tends to rise when the mike is opened. Except for nerves, there's no logical reason for this, yet it happens all the time. Stage actors don't necessarily make good mike performers because they often assume that the energy they need to project across the footlights into a theatre is the same as that required to project into a microphone—and that's not so.

Learn to be aware of and to modulate (vary) your loudness.

In a given environment, we tend to speak within a single, limited loudness range, as long as the overall loudness of that environment remains fairly constant. Recognize this fact, and also recognize that the use of *dynamics* (loudness levels) is one of the cornerstones of drama. A great actor knows when to whisper and when to bellow. You need to know when, too. Listen to people who use their voices to motivate others—athletic coaches, politicians (a few of them, anyway), preachers, salespersons, screen actors—and observe how they vary their dynamics. Also, as was the case with pitch modulation, a little goes a long way. The human ear is very sensitive to changes in loudness, so find the words or phrases in your copy that are the most important, and modify your volume (up or down—you choose) when you read them.

Because you're using a sensitive electronic device, you have a decided advantage over the stage actor. Without amplification, he or she can't whisper and be heard by a theater full of people; you can. Actors have to utilize the loudest portions of their range to portray many intense emotions, such as anger and fear; you don't have to. On the other side of the coin, however, actors are able to use their hands, faces, and entire bodies to get their message across;

you, however, must accomplish the same thing using only your voice. The good news is that that's all you really need: your voice and the listener's imagination.

Pace

Listen to children read. They set a reading speed for themselves and adhere to it, as if guided by some internal metronome, varying only if they stumble over a word or lose their place. Unfortunately, the same phenomenon is readily observable in beginning announcers. I feel that pace variation plays a greater role in holding the attention of an audience than either pitch or loudness. If you have the opportunity, listen to popular radio personality Paul Harvey. Whether or not you like what he says, he's a master of pace variation, speaking at times in a rapid, staccato style, and at other times slowing to a veritable crawl. Or a dead stop. No one uses pauses like Paul Harvey. A pause, correctly used, can be of immeasurable dramatic value. Sometimes Harvey's pauses are so long they border (or go over the border) on the melodramatic, but their effectiveness can't be denied. He does have a very deep, resonant voice, but it's his pace and timing that make him a superb storyteller—and a very highly paid commercial voice talent.

Vary your reading pace.

Whereas pitch and loudness variations should be relatively small, pace variations can run the gamut, depending on how dramatic you want to sound. Good announcers have an inner sense, which tells them when to speed up or to slow down their reading. To develop this sense of timing and pace, you should

1. study the professionals
2. practice

Emphasis

Emphasis (sometimes called *interpretation*) is simply the accenting or stressing of certain words or sounds, to make them more noticeable. Take a simple sentence like "I'm going home tomorrow."

This sentence can be read a number of ways, depending on what it's being said in response to.

1. Is anyone going home tomorrow?
   I'm going home tomorrow.
2. Go home tomorrow, will you!
   I'm *going* home tomorrow.
3. Where are you going tomorrow?
   I'm going *home* tomorrow.

4. When are you going home?
   I'm going home *tomorrow*.

Note that in each case, even though the words are identical, the four sentences convey very different meanings. By shifting the emphasis, these four words can be made to convey four distinct thoughts. Emphasize the important words in your copy.

In a piece of copy, commercial or otherwise, obviously the words you want to emphasize are the ones that will convey the message you desire. Determining which words are to be stressed is easy. With pencil in hand, read the copy *aloud* and listen carefully. Your voice will naturally accent certain words just the way it naturally places an accent on at least one syllable in every word. As you read, underline the words your voice naturally accents. Although your first impression may usually be reliable, don't be afraid to shift the emphasis around within a phrase, and pick the variation that sounds best. Don't forget to read aloud—you need to actually hear the accents—and don't forget to mark your copy.

If you've received a piece of copy from a client who's in the studio with you, it's a good idea to ask permission to mark the copy. The client will never say "no" and will probably be impressed with your courtesy. Now read the copy back, and put a *tiny bit more* emphasis on the underlined words. Again, easy does it! Just lean a little on those underlined words, don't punch them. You'll be amazed at how much better the copy will *sell*. Try this. Before you do any underlining, read a piece of copy cold, that is, without any rehearsal, and record your read. Then go through and mark up the copy, and record another take with the emphases you've decided on. When you compare the two takes, the difference should be readily apparent. The second version will have more life, more authority, more attraction.

Using the Four Variables

There you have them, the four variables that differentiate the good voice talent from the mediocre. Stretch your pitch range, modulate your loudness, vary your pace, and emphasize important words. And I'm not trying to be facetious when I say that your success will depend on three things: Practice—Practice—Practice. Read newspapers aloud; read magazines aloud; read the back of your cereal box aloud; and listen. Listen critically and analytically to the professionals. Don't go to the bathroom or the refrigerator during radio or TV commercial breaks anymore. Listen to how the good announcers *and the bad ones* use their voices. Although it's laudable to learn from your own mistakes, I find it infinitely more productive and less painful to learn from the mistakes of others. Listen, learn, and, above all, practice—practice—practice.

Voice Cadences

Although we don't perceive our speech as musical, the sounds we produce while speaking do, in fact, have pitch. There is a line from the musical *The Music Man* in which Professor Harold Hill remarks to a reluctant singer that singing is merely sustained speech. A bit simplistic perhaps, but true, and because our speech has pitch, it's not unusual for us to speak in regular tonal patterns—musical patterns! These patterns are called *cadences,* and we all have them. The problem is that most of us, without knowing it, use only one (or occasionally two) of these cadences when we find ourselves reading aloud in an unnatural setting, such as before a microphone or in front of an audience. Listen to a beginning announcer, and note how every sentence is read identically. If you listen closely, you'll hear that most of the sentences begin on the same pitch, end on the same pitch, and wend their way through identical tonal patterns. The result is monotony for the audience.

You cannot and should not try to eliminate cadences. They make your speech interesting. What you need to work on is eliminating the constant *repetition* of a *single* cadence. Like a good boxer, you need to mix your shots. Here are three suggestions:

1. Vary the lengths of your sentences. You're less likely to repeat a cadence if your sentences are of many different lengths. Besides, if all your sentences are nearly identical in length, you can't avoid a singsong feeling to your read.

2. Vary your pace and your emphasis. You learned the importance of these two voice variables earlier. A cadence is not likely to repeat if the pace and emphasis change substantially from sentence to sentence.

3. Vary your starting and ending pitches. A particular cadence will always start in the same part of your pitch register—upper, middle, or lower. If you intentionally begin a sentence in a different part of your register, chances are good that an entirely different tonal pattern will follow. The same holds true for endings. Be aware of where in your register you tend to finish sentences, and consciously move your voice elsewhere. If one sentence ends in the lower part of your range, as you read the next sentence, and approach the end, guide your voice naturally into the middle or upper region.

Remember that voice cadences are perfectly natural and desireable; it's the constant repetition of a single cadence that'll drive your audience crazy.

Grammar, Pronunciation, and Enunciation

I must admit a particular affection for the English language, not merely because it's the only language I use, but also because so much good stuff is written in it: the Beatles, Keats, Stephen King, Tom Lehrer (ask for help if you don't know), Shakespeare, Steinbeck. Perhaps I'm being overly chauvinistic about my native language, but it galls me when people wipe their verbal feet on it. It should go without saying that proper grammar is the order of the day in anything you write or produce. Just because a commercial may strive to be friendly or to convey a just-plain-folks feeling, there's no need to pillory the language. Proper grammar *never* indicates snobbery or elitism, whereas, improper grammar *does* indicate lack of education, false modesty, and disregard for your audience's intelligence. Remember also that an audience tends to automatically imbue good media voices with credibility. That's an incredible benefit for you as a communicator, but nothing will pop that credibility balloon faster than sloppy grammar.

Proper pronunciation and enunciation, like good grammar, are indispensible to anyone in radio or TV (or anywhere else, for that matter). *Pronunciation* is the use of the proper sounds in a word; *enunciation* is the process of speaking clearly. Saying "nucular" instead of "nuclear" or giving the second month as "Febuary" instead of "February," or saying "acrosst," "heighth," and "athalete" instead of "across," "height," and "athlete" are examples of mispronunciation. Grammar handbooks and public speaking guides generally have lists of frequently mispronounced words. Check your pronunciation against one or more of these lists, and see how you fare. One sure sign of a beginning announcer is the dropping of the final "g" in words ending with "-ing." Why they do this, I don't know. Maybe they think it makes them sound more friendly, when in reality they merely sound illiterate. Off the air, feel free to speak any way you choose; but on the air, proper pronunciation is a must! Correcting your pronunciation of a few words isn't tough; unfortunately, though, you probably won't receive many compliments. Pronunciation is one of those things that generally goes unnoticed—until you slip up.

One final note on pronunciation. People from different areas of the country (or the world) pronounce words differently. In Massachusetts, where I hail from, we practice linguistic recycling—we take r's from "pahk" and "cah" (*park, car*) and stick them on the ends of "Chinar" and "Cubar." A regional accent is no excuse for sloppy pronunciation. I remember watching the news from a Georgia TV station many years ago and thinking how remarkable it was that the newspeople spoke with only a hint of what I perceived as an accent. They could have been broadcasting from just about anywhere. If you mispronounce words, blaming your accent is just an excuse. Just as you can learn to speak *with* an accent (as actors frequently are called upon to do), so you can also learn to eradicate or at least minimize an accent. You may need professional speech coaching to help you, but it's worth the time and money if you plan to talk for a living.

Enunciation is clarity of articulation, making sure your pronunciation is clear and distinct, not merely correct. My mother used to tell me that people were being *lip-lazy* when "four score and seven years ago," would come out "fourscorin' sem yiz ago." Enunciation is merely the correct use of your lips, teeth, and tongue in the formation of words. Correct pronunciation is for naught if your words come out slurred and sloppy. These slurs can happen to anyone who ever tried to race through "Peter Piper picked a peck of pickled peppers." Sometimes you'll hit a word or a combination of words that you can't seem to get your tongue around. My personal bugaboo is consecutive words beginning with *th-*. A phrase like "the thick thing" gives me fits. There are, fortunately, a few ways to get around problems of enunciation:

1. Slow down! Usually when your tongue trips over a word or phrase, you're speaking at a pretty fast rate. Diminishing the rate may give the lips, teeth, and tongue time to adjust to the problem word(s).

2. Emphasize a problem word. If I say "the *thick* thing," leaning on (and thus slightly lengthening) the word *thick*, I can get through the phrase. Try this technique with phrases that give you trouble. An alternative is to try breaking up the phrase with an additional word, as in "the really thick thing," or "the thick slimy thing."

3. If all else fails, substitute synonymous words for the ones giving you trouble. I was producing a commercial for an announcer who couldn't spit out the phrase "sheep shearers," so we used "sheep clippers." "Rubber baby buggy bumpers" would be easier to say as "rubberized baby carriage bumpers"—easier, but not as much fun.

Not to scare you off, but changing the speech patterns you've used all your life is no easy chore. In fact, it's probably the toughest aspect of production to master; it's certainly the most time consuming. Learning to master a piece of equipment may take hours, days, or weeks, but learning to master your own voice takes years. Fortunately, to practice, you don't have to go anywhere in particular or use any special tools, although it's wise to record yourself and listen critically on playback. A friend of mine who is a marathon runner told me once that running 26.2 miles isn't easy—but it's very simple. Just keep putting one foot in front of the other, and eventually you'll finish. In other words, don't look at the distance as a huge chunk; see it as a series of little chunks. This is good advice to follow, whether you're running a marathon or learning how to use your voice.

If, in everyday life, people judge you by the way you speak, then when your voice is on the air or in a commercial, your use of language must be impeccable. There's nothing arrogant or stuffy about correct grammar, pronunciation, and enunciation. The fact that this notion exists may be a troubling indication that declining language skills are becoming the norm rather than the exception. We in the electronic media wield enormous influence through our use of language; therefore we must hold ourselves to the highest possible linguistic standards.

Review

1. Any voice can be trained to be more effective, but the process is difficult and time consuming.
2. Practice will not change the natural pitch range and harmonics of your voice.
3. Four variables are involved in voice training: pitch, loudness, pace, and emphasis.
4. Learn to stretch your natural frequency range slightly.
5. Be aware of and vary your loudness (dynamic range).
6. Vary your pace. Don't merely find a comfortable pace and stay there forever.
7. Shifting emphasis can result in a sentence or phrase having a number of possible meanings. Determine the important words throughout your copy and slightly stress these words when reading.
8. Learn to vary your voice cadences. This can be accomplished by varying the length of your sentences, by changing your pace and emphasis, and by varying the pitch at which you begin and end each sentence.
9. Poor grammar, pronunciation, and enunciation don't make you sound friendly; they make you sound uneducated.
10. *Pronunciation* is the use of the correct vowel and consonant sounds in a word. *Enunciation* is the clarity of your speech.
11. A regional dialect is no excuse for faulty speech.
12. Correct enunciation can be aided by slowing the pace, by emphasizing a particularly difficult word, or by breaking up a tongue-twisting phrase with an additional word or two.

9

Mixing

The best producers are those who can combine separate elements into an organic whole that has a greater impact on the listener than any of the individual parts. Although a good ear is paramount when mixing, good ears *are developed* through adherence to sensible production procedures.

Setting Levels

Before beginning *any* production, do the following:

1. Clean the tape recorder heads, guides, capstans, pinch rollers—anything that comes into contact with the recording side of the tape.

2. Set your recording levels. See Chapter 2, "The Console," for the procedure for doing this. Your reference tone can originate from any one of four sources: the console, an individual tape deck, an external tone generator (see Figure 9.1), or your own voice. Obviously, an electronically generated tone, because it can maintain a preset frequency and amplitude without any fluctuation, is preferred. Many multitrack consoles and multitrack tape decks come with a built-in reference tone (usually 1 kHz; see Figure 9.2). In the absence of such a feature, you should have a tone generator (see your engineer) at your disposal, which you can hook into your system (generally patched into a console input via the patch panel). If none of these options is available to you, as a last resort get on mike, hum a steady pitch, adjust the console so that your voice reads 0 Vu on the console meter (or as close as you can get), and without moving off mike (move the mike if you have to), note the position of the tape deck meter. Adjust this meter up or down so that it's aligned with the console Vu. It's a good idea to repeat the procedure once or twice just to make sure the meters are aligned. If there are other meters on other pieces of equipment to be set, you can either use your hummed tone, or use the tape deck meter you just set as your reference. Talk into the microphone and observe how the other

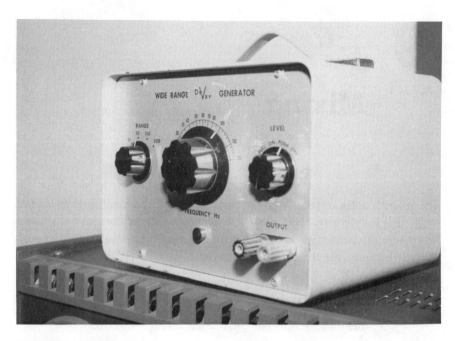

9.1 Small tone generator.

meters act, compared to the first tape deck meter you aligned; adjust the other meters so that they move just like that first tape deck meter moves.

3. If possible, set playback levels. Do your tape decks have playback volume controls? If not, skip this section. Playback will be controlled entirely by the console. If your decks do have such controls, set them in this fashion:
   a. Make sure record levels have been set properly.
   b. Thread a tape on the deck, and roll the tape in the record mode, thus recording your reference tone at 0 Vu (console and deck meters).
   c. With the tape rolling, switch the deck monitor select to playback (or repro). Even though the tone is being recorded, the deck's Vu meter(s) are showing the level of signal *already imprinted on the tape*—that is, the playback level. Using your playback volume controls, make the meter read 0 Vu. Now the deck's record and playback levels are identical.

Some decks come with a dual playback level control feature, allowing you to either set your own playback level, or to use a preset playback level built into the deck at the factory. Although I've found factory-set playback levels to be perfectly acceptable, I don't like to surrender control of any aspect of production to someone else. I know engineers who, after setting a deck's levels,

9.2 Internal tone generator on a tape deck.

remove the control knobs, figuring that since the levels have been set perfectly, why would anyone want to adjust them further. This is a fine theory—but then again, in theory, the Titanic couldn't sink. What happens if you have to play a tape that was recorded somewhere else, and you find that it was recorded at too low a level? If your playback level control is preset in some fashion, you'll be unable to play this tape back at an acceptable level, except by cranking it up through the board. In general, the rule is this: *Don't relinquish any control of any aspect of production to anyone else.*

Cleaning your gear and setting your levels should be standard operating procedure before you start work. The whole process shouldn't take more than 2 or 3 minutes, minutes that can save you hours of wasted energy and frustration.

Foreground/Background Interaction

The most important element of an audio commercial or promo is almost always the voice. Therefore, the voice is almost always produced first. To supplement the voice, and to provide greater impact, the producer may add any or all of the following:

1. sound effects
2. music
3. special effects

Important point: Unless your voice track is perfect (or as close as you can honestly come), don't bother to go on to any other aspect of the production. All the music and effects in the world won't camouflage a sloppy voice track!

If you intend to make good use of these elements, you must visualize your production as consisting of two distinct layers—*foreground* (voice or whatever the attention is supposed to be focused on) and *background* (supporting material—music, sound effects, and special effects). When these two layers work together, the sound is aesthetically pleasing, and, consequently, the message is more apt to be psychologically effective.

The concept of foreground/background interaction is largely ignored or unknown in radio production today. However, just as ignorance of a law is no defense, ignorance of a production principle is no excuse. Ignoring a technique that will make your work better is tantamount to putting your stamp of approval on mediocrity.

Because most radio production is done by the announcing staff, the announcer generally does his or her production prior to or just after an on-air shift. Production is looked upon as a burden to be borne, and so most of it consists of voice with a music track slapped behind it. In cases like this, the foreground and the background don't really interact. True, the music supplements the voice, but the two planes, foreground and background, remain completely separate.

To understand foreground/background interaction, you must be aware of two all-important principles:

1. The ear, like the eye, follows movement.
2. The ear likes to hear movement.

Let me give you an example. Imagine that you're equidistant from two speakers, like the apex of a triangle. From the left speaker you hear a steady 100-Hz tone, and from the right speaker you simultaneously hear a single sound effect that shows action of some sort—footsteps, a car engine starting, a player piano at work, a dog barking, etc. If the sound levels from the two speakers are identical, your attention will be drawn toward the moving sound. In fact, even if the steady tone is noticeably louder than the sound effect, your mind will still focus toward the right speaker. Also, the duration of the sound effect is of no consequence. If a single gunshot were to be heard from the right speaker,

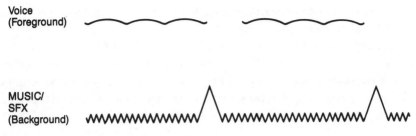

9.3 Foreground/background interaction.

our attention would instantly shift from the left to the right, and, as the shot quickly decayed, back to the steady tone on the left.

In most commercials, unless the producer deliberately intended otherwise, you're not supposed to notice the background. If you do, then you're not paying attention to the message in the foreground. You can't do both easily, any more than you can seriously listen to two people talking at the same time. The mind can focus on only one incoming line of information. However (and this is the fundamental principle behind foreground/background interaction), when there's a break in that primary line of information, even a small break, it's perfectly acceptable for a *secondary* line of information to assert itself and, for the duration of the break, come to the foreground.

This, in essence, is foreground/background interaction (see Figure 9.3). When there's a pause in the commercial message, whether between words, phrases, or sentences, for that short time, our attention begins to shift toward what's moving—that is, the background (music, sound effects). If we boost the level of the background *slightly* during the pause, the background will momentarily become the foreground. When the voice reenters, drop the music and/or sound effects back to their previous levels. The result is this: the ear was focused on the voice. When the voice paused, the background came up, causing the ear to shift its focus. When the voice reappeared, the background dropped back, and the ear shifted back to the voice. This manipulation of the mind's ear is what foreground/background interaction is all about.

When we shift the listener's focus, we hold the listener's attention!

However, this is not to say that you can knock the listener's focus back and forth like a ping pong ball. If you sharply boost the music bed at the end of every sentence, your work will sound choppy, mechanical, and extremely annoying. Look for breaks in the copy. Most commercials have breaks, places where the text changes tone, style, or direction. These breaks usually provide splendid (and logical) opportunities to make the background do something other than lie there behind the voice. A *small* boost is sufficient to bring the background to the fore. Remember, the ear will follow *any* movement that it perceives, and a little movement goes a long way.

Remember also that you only want to shift the focus for a very short time, until the voice comes back. If you blast the background up through a short pause, the listener will have a tendency to focus too much on the background and thus may resist the shift back to the voice. Always keep in mind that these shifts are analogous to gentle nudges, not violent shoves.

Background Music

Two factors should govern your choice of music in a piece of production:

1. need
2. suitability

There's no law that says a production must contain music, although some producers, whether out of habit, lack of confidence in the voice, or simple knee-jerk reflex, seem to feel that their work won't make it without a music bed. On the other hand, after a 4- to 6-hour air shift, the average announcer/producer justifiably has one thing in mind—departure. So he or she might grab the assigned production, head into the studio, lay down and cart up voice tracks, and break for the door. I know of no pearls of wisdom to employ concerning the selection (or nonselection) of music. I used to think that the more serious or somber the copy, the less the need for music: conversely, the more upbeat the copy, the greater the need for musical backing. However I've found too many exceptions to this idea to be able to set it in stone, so use it only as a reference, not as a rule.

A music bed's suitability would, I hope, be a matter of common sense. Because the music is supposed to supplement the copy, the music should *reflect* the copy, especially in terms of tone and tempo. If the read is bright and upbeat, the music should be, too. You're not trying to fit the music only to the product being advertised: let the *voice* reflect the *product*, and let the *music* reflect *both*. Sometimes, a particular type of business does call for a specific type of music. Kiddie music (toy piano, chimes, flute) would nearly always be fine for a toy store, or how about "Hooray for Hollywood"-type music for a video rental business. Other businesses have less obvious mates; for example, what kind of music would you use to bed a spot for a hardware store, or a lumber yard, or a supermarket, or a clothing store? I once spent an hour rummaging through dozens of discs with a client who wanted *some good indoor-outdoor carpet music*. Forget it! If the product or business doesn't instantly lend itself to a specific type of music, look for music that fits the voice!

Also, never, *never* use a song or anything with a lyric to back your production. Remember the example of trying to listen to two people talking simultaneously? That muddle is what you'll get if you have someone singing behind your voice track. In order for this other voice not to be a distraction, you'd have to lower its volume so much that it would cease to have any value to you. If the singer is providing a backup vocal (oohs, ahs, and the like) that's different. In general, though, you'll have a tough time holding an audience's attention if there's someone singing behind you.

Music Analysis

Once you've decided on a particular cut of music, I suggest you do something that very few producers do—listen to the music from beginning to end. Why? Musical punctuators! Punctuators are short, sharp sounds that instantly, though momentarily, capture our attention, sounds like a gunshot, shattering glass, a scream, or a slamming door. Music can have punctuators too, elements that tend to stand out from the rest of the composition. A music punctuator can be a drum hit, a key change, the entry of a new instrument—even abrupt silence after a particularly raucous passage can act as a punctuator. For example, on the Young Rascals' 1966 hit "Good Lovin'," near the end, as the band is roaring along, they hold one note at full volume for a couple of seconds—and then instantaneous silence. For nearly two measures there's nothing, and then they kick back in at full volume again. How about the end of "A Day in the Life," the final cut on the Beatles' *Sgt. Pepper's Lonely Hearts Club Band*? The music slowly builds to a deafening crescendo, abruptly stops, and then crashes back in with that seemingly interminable final chord. Instant silence can be a *very* effective punctuator. Are there any punctuators in your music bed? If so, you have two choices:

1. ignore them
2. use them

Producers in a rush will undoubtedly choose the former. Don't do it! You have no control over what's in the groove of a disc: so when punctuators appear in a cut of music, look on it as a gift, a bonus, a serendipitous event—a way for you to make your production shine. This is the perfect time for some foreground/background interaction. What you want to do is to *cause a punctuator to sound when there's a pause in the copy!*

Time your music, note when the punctuators sound, and adjust or rewrite the copy (with the client's permission, if necessary) so that the pauses in the voice and the punctuators in the music align—once or twice during a :30 spot will be enough to enhance your reputation. It will take extra effort and patience to get the timing down, but the result will be well worth the effort. How often on "The David Letterman Show" have we seen short (30 seconds or less) pieces of business that took many hours of work and hundreds of dollars to produce? The point here is simple:

QUALITY TAKES TIME.

However, let's be realistic. If, after a full day on the air, you have a dozen pieces of production to do for tomorrow morning, you're not going to spend a ton of time on each one. Pick a few, and do something special with them, anything to make your production other than ordinary.

Multiple Music Tracks

Just as there's no law requiring you to use music, there's also no law requiring you to use only one music track per spot. A music shift can make for a very effective punctuator if

1. the pieces of music are completely different in tempo, tone, and instrumentation
2. at some point, the copy clearly shifts its focus or tone

However, beware! You can't just butt two completely different pieces of music up against each other without severely jarring your audience. If you want to use more than one piece of music in a single short piece of production, be prepared to do one of the following:

1. Put at least :02 between the end of one piece of music and the start of the second
2. If the pieces of music must be tight together, cover the juncture with a loud punctuator (cymbal crash, bomb explosion, or the like). Under cover of the punctuator, dump the first piece of music out, and before the punctuator has fully decayed, start up the second piece.
3. If the two pieces are in the same key, you can sometimes edit the two together. The similarity of the keys can often cushion the other differences. See Chapter 10, on "Editing," for the correct procedure.

Using different pieces of music can reinforce separate sections of your copy. It's quite remarkable how much more pizazz your work can generate with judicious and correct use of simple tools. It's equally remarkable how few producers either know of their existence or bother to take the time to use them.

Fade Out versus Cold End

All music tracks end in one of two ways:

1. *fade out,* when the music continues as the volume drops
2. *cold end,* when the piece ends on a single note or chord

I must confess a personal preference for a cold ending. It has a sense of finality that a fade out doesn't, and it can, depending on how abruptly that final note ends, provide you with a final punctuator. Mind you, there's nothing wrong with a fade out. It makes for smooth continuity on the air (as the music fades, bring the next piece of business up, so there's a bit of overlap). Also the fade-out ending doesn't require the timing that a cold end does. As soon as the voice is through, immediately fade out quickly and smoothly. A second or two of fade is *plenty.* If the fade goes on for too long (I've heard beginners take :10 or more), the audience doesn't know if perhaps there's more commercial coming, or if this is a record that the announcer played under the commercial. For a listener, there aren't too many things more annoying than confusion. A cold end can go on somewhat longer than a fade (:02–:04 if necessary), because the listeners can hear the music coming to a close. They can sense that the message is over and that the music they're hearing is about to end.

If you're using a cold end, don't have the voice and the music end together. That final note acts as a ribbon, tying the spot together with a solid ending. However, that final note must be heard in the clear to be effective. If it's under the last word of the voice, the effect is lost.

Mixing Voice and Music

Combining voice and music is easy to do but difficult to explain. There are no hard and fast rules, because *you* mix according to the dictates of *your* ear. However, you do have to bear two things in mind:

1. the voice and music levels relative to each other
2. the overall level of the mix

Here's my procedure. First I bring up the voice and set it to about *80%* on the console Vu meter. If I set my voice track to 100%, and then added the music, obviously the overall level would be well into the red. So back the voice off a bit. Then bring up the music under the voice, until the music begins to mask or interfere with the voice. Then, pull the music back a bit. The overall level of the mix should be fine, and the voice should be perfectly intelligible.

When you play the mix back, don't follow it on your copy. *Listen to the mix as if you're hearing those words for the first time.* Is every word distinct?

Does the music support the voice or compete with it? Did you remember to slightly boost the music when the voice paused? The average producer would just set the voice and music levels and let them roll, but who wants to be average?

If everything's the way it should be, then mix down onto a cart or whatever format you use. Play the cart back, and listen critically to the final product. If it's good, fine. If not, don't be afraid to remix. It's always a good idea to write down the console level settings you have used. That way, if you have to remix, you don't have to waste time resetting your voice and music levels from scratch.

One other thing. Don't use headphones when setting your levels or mixing. Most of your listeners will probably hear your work through a speaker, so mix using a speaker as your monitoring source.

If you're going to use any kind of music, the music will probably (but not always) begin just before the voice (:01–:01.5) and end just after the voice. Thus, if you're planning to stay within :30, for example, you're going to have to allot some time for your music intro and the music close. I recommend that you aim for a voice track that's :27–:27.5 long. That gives you time for 1 second of music on both ends. Most of the production studios with which I'm familiar ask the announcer to shoot for :27.5, especially if it's a TV spot. TV stations are notorious for chopping the end of a spot that runs over :29.5 or :59.5, and my hat's off to them!

Watch the clock

If a client has purchased time for an announcement of a specific length (:30 or :60, generally), it's your responsibility to ensure that the spot doesn't run over that time limit. I know a lot of stations where they let a spot run a second or two (or more) over the limit without thinking of the consequences. All you need is for Client A to come to you complaining that his spot is :30, but that his competitor, Client B, has a spot running :33. Do you give Client A extra airtime to even things out? Do you produce a spot for Client A that's :34 or :35 to make him feel better? Mark Twain said "If you always tell the truth, you'll never have to remember what you said." If you keep all your production within the allotted time limits, you'll never have to remember whose commercial is how long.

More of my students lose more credit for time violations than for any other single infraction. Producer beware! Also, don't think that the timer stops when the voice is over. If your voice track ends at :29, but the music slowly fades out for another :05, you've got a :34 spot! The clock stops when there's no more sound.

Music Sources

So where do you find good production music? There are a number of really good music libraries around (see Figure 9.4), and every radio station should have one (or more). The trade publications always carry advertisements

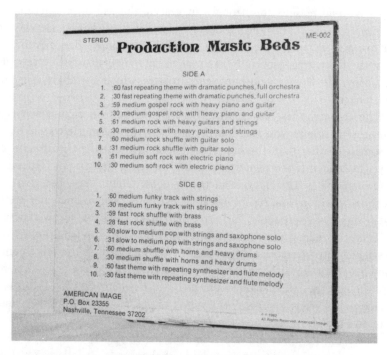

9.4 Disc from a production music library. (Courtesy of American Image.)

from production companies with libraries to sell or rent. Don't overlook recommendations from other producers. They're generally reliable and unbiased. If you attend any of the radio broadcasting trade shows held every year, you'll probably find a number of production companies pitching their music and/or SFX libraries.

As you might guess, libraries differ greatly in scope and content. Some offer :30 and :60 music beds, but don't be surprised if the times indicated are a second or so off. Others supply you with longer beds, which you can customize to any length you want (see Chapter 10, on "Editing"). Some provide beds with intermittent singing by professional jingle singers, allowing you to sandwich your copy between jingle segments and to produce (so the library producer would have you believe) a more professional spot.

Some libraries are available for outright purchase, whereas others you have to lease for a set length of time (frequently 1–2 years). The price will vary but will invariably be high. It's not unusual for leases to run to many thousands of dollars per year. Some libraries, especially those used by film and video producers, require you to pay by the needle-drop. In other words, you pay each time you use any of their compositions. However, because radio stations use lots of music, libraries geared to their needs rarely charge by the needle-drop.

I have two reasons for strongly suggesting the use of production music libraries. First, the quality and variety of music is likely to be considerably better than average. Second, you won't have to worry about copyright violations. This is a very sticky area, especially as far as clients are concerned. Although I'm not a copyright lawyer, I do know certain facts, which I strongly urge you to take to heart.

The copyright law prohibits unlicensed duplication of *any* copyrighted material. Put simply, you may not use any copyrighted music to make money for someone other than the copyright holder without the permission of the copyright holder. If the owner of a pet store wants you to put "Hound Dog" or "How Much Is That Doggie in the Window" behind his spot, *don't do it!* These compositions and recordings are copyrighted, and you may not use them to sell your client's products. To do so would make you liable for two lawsuits— one from the federal government, and one from the individual holder of the copyright. Getting permission is absurdly complicated, as the Nike company found out when they used the Beatles' "Revolution" in one of their commercials. Nike thought they were covered, but *no-o-o-o*. The copyright for a song's lyrics may be held by one company, the rights to the melody by another, and the rights to a specific recording by someone else.

The penalties (fines and imprisonment) are extremely steep, especially the fines, which can run into the hundreds of thousands of dollars. Yet, I doubt whether there's a market in the country in which there's not at least one station illegally using copyrighted music. This, of course, makes things tough for you, when your pet store client says that WZZZ across town didn't hesitate to use "Hound Dog." You don't want to lose the business, but are you willing to put your station at risk? Although most violators are not caught, some are. It's not unusual for a station that adheres to the law to report competing stations who are copyright cheats to the FCC. Are you willing to take the risk? As Clint Eastwood once remarked, "Do you feel lucky?" Don't give in. You won't endear yourself to the client or to the salesperson who worked to get that client to place a buy with your station, but there comes a time when integrity has to take precedence over profit, and this is that time.

There are two exceptions to all this brouhaha. If the holder of the copyright directly benefits from the use of that music, then you're probably safe. If, for example, a singer is going to give a concert in your area, it's permissible to use his or her recordings in the commercials promoting the concert, because it's the singer, not some pirate, who benefits by the music's use.

Also, copyrighted music may be used to promote your radio station, or in events promoted by your station—*if* you have paid the requisite fees to ASCAP and BMI, the two major music licensing organizations in this country. Because the licenses to almost all popular music are held by these two bodies, radio stations, in order to be licensed to play this music over the air, must pay annual fees to ASCAP and BMI. That's right, your station has to pay to play. The fees, which go to the artists, composers, and producers, are very hefty. Payment of those fees not only entitles your announcers to play this copyrighted music as much as they want, but it also entitles your station to use this music in announcements promoting the station. Although the pet store owner couldn't use "Hound Dog" to sell puppies, your station can use it to promote a station-sponsored Ugly Dog Contest.

Not all music is copyrighted. Many standards are in the *public domain*, which means that no one receives any royalties from this music's use, and, consequently, it's fair game for anyone who wants to use it. "Happy Birthday to You" is in the public domain. So are "Auld Lange Syne" and "Yankee Doodle." Not all the music in the public domain is ancient. Any music can pass into the public domain if the composer dies, and fifty years pass without the copyright being renewed.

People have tried to come up with a number of dodges designed to circumvent the law. One client told me that as long as I didn't use more than six measures of the music, I was safe. Another said that as long as I didn't use a part of a song in which the title was sung, I was in no jeopardy, and there are other machinations too ludicrous to mention. So remember, if you're using music in commercial announcements (except in the two special cases mentioned earlier), the only way of avoiding copyright infringement is to use either music from the public domain or music from a reputable production music library.

Here's a suggestion. For your station promos (or other noncommercial pieces), try using music from movie soundtracks. It's generally high quality, well produced, and designed specifically to enhance particular moods. Also, it may be familiar to your listeners, and familiar music makes a listener feel comfortable (witness the success of radio stations with *Oldies* formats). "But aren't movie soundtracks licensed by ASCAP and BMI?" I hear you ask. Yes, they are, but note that I'm suggesting you use these pieces of music for station promos. I'm assuming your station has paid its licensing fees. If not, movie soundtracks are also off limits.

Sound Effects

Nothing can make your production as realistic, as believable, or as exciting as can properly applied sound effects (SFX). Whereas music can make a spot pleasant and appealing to our aesthetic sense, SFX can make your work come alive. Music coaxes listeners into the commercial—SFX pull them in bodily.

Although the number and variety of sounds available to us is almost infinite, I've found it handy and simple to use only two broad classifications:

1. *punctuators*—short, sharp, generally high-level sounds, designed to attract immediate attention
2. *ambient sound*—environments, generic background noise, designed to create a mental image

Punctuators

SFX punctuators operate like music punctuators. They instantly grab your attention but hold it for a very short time, sometimes less than 1 second. Punctuators can be separated into two categories:

1. sounds that reinforce action within the production
2. sounds that suggest some sort of undefined action

For example, let's take the sound of a shattering glass. If our commercial depicts someone eating a meal, and the person says "Where's my glass? . . . *oops*" (shatter), the sound reinforces or reflects the action. However, let's say we insert that same sound into the opening of "The David Letterman Show" and create something like "Paul Schaeffer and the World's Most *dangerous* Band" (shatter). Here, the sound doesn't suggest anything definite. Some ephemeral connection has been made between the concept of "dangerous" and the sound of the shattering glass. It is left entirely up to the listener to decide what that connection is.

Whenever you use a punctuator as a reinforcement, the sound has to make sense, and it must be *instantly* identifiable. If a character in a spot drops a glass, you don't expect to hear a bomb explode. Although we focus on and off a punctuator very quickly, if the punctuator is muffled, muddy, or otherwise hard to identify, our minds will stay with the sound, trying to identify it, even to the extent of temporarily ignoring the voice.

A punctuator used as a suggestion of undefined action can be almost anything. Logic doesn't necessarily play a part here. Because you're not restricted by reality or the appearance of reality, anything goes.

Although you'll probably get most of your SFX from discs, don't hesitate to create your own. One of my students wanted the sound of a man running through the woods. He brought in a large box filled with leaves, set a mike down by the box, climbed in, and ran in place while reading his lines. The sound was perfect. One year, our production students produced a play that called for a character to ostensibly type on a computer keyboard the lines he spoke. We had computer keyboard sounds in our sound library, but none that would synchronize well with the actor's lines. The students came up with a number of clever and usable solutions. One tapped on a table with two inverted Dixie Cups. Another pushed the tape control buttons on an unused tape deck. Another simply brought in a computer keyboard. In all these cases, the students synchronized their keyboard tappings with the prerecorded voice track, and the result was always convincing.

If you make your own sounds, bear in mind that your creation doesn't have to be a perfect copy of the sound you're trying to create. All you have to do is come close, and the listener's mind, moved along by the power of suggestion, will do the rest for you. As long as the audience knows what the sound is supposed to be, you're safe. If you suggest to your audience that you're going to ring a bell, and you whack an old automobile brake drum (available at your local salvage yard) with a metal rod, their ears will hear, and their minds will *see* a bell. That's the beauty, the absolute magic of sound: the ability to create convincing and sometimes complicated illusions in people's minds, using mostly mundane, everyday tools.

Ambient Sound

Whereas punctuators are designed to attract a listener's attention, ambient sound is supposed to be an unobtrusive presence—the background. An ambient sound track can give life to a production better than anything else I know of. A dry (nothing else added) voice can effectively laud the merits of a particular restaurant or auto mechanic or department store, but if you add an appropriate ambient soundtrack, the credibility, and thus the effectiveness of the piece, will increase astronomically.

Just as you listened to your music track for any punctuators that might be present, so too should you check your ambient soundtrack for sound punctuators. There may not be any. However, if there are one or two, don't hesitate to adjust your copy, as you did with the music, to make use of punctuators that come your way by chance. A restaurant track may have a noticeable clink of utensils. In your garage track, you may hear some noisy machine switch on. A cash register bell in your department store track may provide a nice punctuator. Keep your ears open. Following is a list of sounds, grouped by type, from the British Broadcasting Corporation Sound Effects Library:

1. airplanes—jets, props, turbo-props, Piper Comanches, helicopters, bombers, flying boats, vertical take-off; take-off, taxi, land, in flight, crash.
2. animals—birds—cockerel, skylark, canary, parakeet, parrot, thrush, nightingale, goose, rook, pigeon, owl, sparrow, starling, swan, crane, etc.
3. animals—domestic farm—chickens, dogs, horses, pigs, cats, sheep, cows, geese, goats, squirrels, rooster crowing thrice, cat fight; growls, howls, yelps, yaps, etc.
4. animals—insects, fish, sea—cricket, bee, wasp, gnat, mosquito, fly; shrimp, cowfish, catfish; alligator, frog, toad, turtle, seal, dolphin, whale.

5. animals—wild—aardvark, elephant, lion, tiger, hippo, sea lion, chimp, hedgehog, bat, bear, badger, polecat, otter, boar, reindeer, fox, wolf, etc.

6. animals—wild—lion, ape, leopard, monkey, panther, orangutan, gibbon, zebra, gnu, vulture, hyena, rhino, lizard, tiger, cobra, adder, wild dogs, etc.

7. backgrounds—interior—cocktail party, hotel, prizefight, hospital, laundromat, restaurant, café, bus station, concert hall, factory, bowling alley, rink, etc.

8. backgrounds—exterior—garden, park, lakeside, pool, city streets, highway race track, parade, playground, football game, building, birds, sea shore, etc.

9. cars and trucks—Le Mans race, Wolseley 1660, Triumph TR3, 1914 Model "T," 5-ton Austin Diesel Truck, taxi—start, stop, idle, shift, pass; horns, etc.

10. cars and trucks—Austin A35 Van, Ford Cortina, Triumph 650cc Twin Motor Cycle, drag race, skids, crashes, tunnel traffic; blower, wipers, etc.

11. children and babies—baby fussing, crying, sobbing, belching, sneezing, laughing, with toys; toddlers thru teen-age in school, playground, playing field.

12. city sounds—fire station, fire engine, fire scene, police siren, whistle, ambulance, crowd, riot, strikers, traffic, escalator, pedestrians, theater, etc.

13. electronic—offbeat—bubblings, whistles, creaks w/wo reverb, crashes, coughs, clocks, booings, doings, squeaks, pumps, pops, sirens, whooshes, etc.

14. farm and country—ploughing, felling, haying, horseshoeing; scythe, axe, chainsaw, pumps; forking hay, tree coming down, wood chopping, etc.

15. farm and country—frog pond, country airport, canoe race, brook, splashes, lakeside, motorboats, dam, mill, stable interior, horse and sleigh, moped, etc.

16. foreign countries—Arabs in desert, Japanese theater and parade, Tibetan monks; London, Greek, Chinese, Dutch, African scenes.

17. guns and warfare—revolver, machine gun, rifle, 18th century naval battle, cannon, torpedoes, depth charges, pom poms, sonar, air strike, marching, etc.

18. guns and warfare—air raid, dog fight, explosions, A-bomb, flying bomb, tanks, pistol, grenade and mortar fire, 21-gun salute, rocket launcher, etc.

19. home sounds—lawn mowers; doors, windows, curtains, locks, bolts, chains, fans, clocks, bells, phones, knocks, buzzes, hums; vacuum, flushes, water

20. home sounds—kettle, percolator, dish washing, water boiling, ice tinkling; sewing machine, camera, blinds, doors, washing machine, cards, etc.
21. machines—vending machines, pinball, cash registers, drill, car wash, saws, oil rig, dentist's drill, compressor, generator, steamroller, etc.
22. people's reactions—laughter, screams, cheers, sobs, applause, coughs, sighs, groans, hiccups, gasps, yawns, burps, sneezes, footsteps, etc.
23. period effects—cavalry calls sequence, horsedrawn traffic, mail coach, paddle steamer, steam railway, monoplane, streetcar, movie organ, etc.
24. public events/places—bus station, train station, stores, amusement park, auction, street festival, restaurant, park, pub, march, parade, etc.
25. religion and rites—Armenian, Catholic, Episcopal, Jewish, Methodist, Hari Krishna, Zen; church bells & interiors; barbershop quartet, taps, reveille, etc.
26. sea and water—seagulls, seawash, diving; liner & tugboat sirens; 40HP outboard, motor launch, rowboat, Hovercraft; marine siren, winch, chain
27. sea and water—sail boats, fog horn, ship's buoy, bell buoy, rowboat, submarine, jet passenger boat, clipper sailing ship, sails, waves, tackles, wind, etc.
28. sports and recreation—tennis, fencing, squash, ice skating, ice hockey, golf, football, soccer, bowling, archery, basketball, cricket, ping pong, skeet, etc.
29. supernatural/space—phantoms, wizards, goblins, monsters; gravity generator, time warp, enchanted forest, flying saucer; twangs, zings, bangs, etc.
30. trains—diesel, steam, electric; passenger, freight; classic trains, Metroliner; starting, running, climbing, arriving; whistle; shunting
31. transitions and cues—pips, bubbles, gongs, cymbals, bells, clocks, ticks, horns; music boxes, barrel organs, string and piano notes, electronic tones, etc.
32. violence and horror—screams, heartbeats, howling; Dr. Jekyll's lab, Dracula, Frankenstein; werewolves, hellhounds, grave digging, the guillotine, etc.
33. weather—thunder, hail, sand- and snowstorms; gales; moaning, eerie winds, thunder clapping, booming, rumbling; rain; etc.
34. weather—wind tunnel, storm at sea, rain in city and country, on metal, wood and concrete; slowed down; thunder; storm; flags flapping, etc.
35. work—factory ambience, whistle, time clock, loading bay, crane, forklift; phones, coffee break; office procedures and machines; etc.

36. combat—karate exercises, breaks and kills; Wild West effects of cowboys and the saloon; axe and swordfights, medieval jousts, armed camp, etc.

37. even more death and horror—staking a vampire and other hard-to-record-by-yourself FX, including throat and wrist cutting, stabbings, various tortures, imaginary and murderous birds, beasts, and reptiles, etc.

38. Dr. Who—numerous intergalactic and extradimensional events on and between seven alien worlds, plus audiograms of the doctor's mind and Tardis operations

39. the electronic home—food processors, smoke alarm, ice maker, microwave oven, garbage disposal, corn popper, sideburn trimmer, more

40. the electronic office—all those new and interconnected communicators that compute, process words, transmit facsimiles, etc.: clicking, beeping, whirring, buzzing, talking to one another

41. holidays: home and abroad—sports and recreation, foreign locales, in- and outdoor fun and games

42. location backgrounds—extended backgrounds: country stream, aerial currents, garden in springtime, seashore, forest adagio, rain

43. medicine—operating room prep, anesthesia, surgical procedures, CAT-scan, kidney dialysis, blood pressure, etc.

44. science fiction—synthesized futurama: the 21st century, including Dr. Who and the Hitchhiker's Guide to the Galaxy

45. silent movie music—comedy, sentiment, drama, action, formal occasions, locations, more.

46. sounds of suburbia—exercise class, shopping mall, cheerleading, gardening, party, supermarket, town meeting, dentist's office, orthodontia, etc.

47. sounds of speed—SST (Concorde), Hovercraft, jetfoil, highspeed train; fire emergency and rescue, police, ambulance, helicopter rescue at sea.

48. urban life—pizza and coffeeshops, doorman, hotel lobby, elevator, garbage truck, street repairs, etc.

49. sports and sporting—badminton, basketball, billards, bowling, boxing and sports of all kinds from skateboarding to pheasant hunting.

50. video games/entertainment—Pac-man, arcade din, Donkey Kong, Asteroids, Galaga, Robotron 2004, disco and birthday parties, etc.

(Reprinted with permission of Films for the Humanities and Sciences, Inc.)

Creating your own ambient sounds can be a tedious chore or great fun, depending on your attitude. Analyze the sound you want. For example, let's say you want to make a restaurant atmosphere. Mix together the following:

1. generic milling crowd
2. quiet music
3. clattering dishes and glasses

Voilá! Instant restaurant. I do this with any ambient sound I have to create. The trick is to hear the sound in your head first and then to analyze what you hear.

Mixing Voice and Sound Effects

As with voice/music mixing, let common sense prevail. Punctuators are meant to come into the foreground, ambient sounds stay in the background. Obviously, a punctuator should, on average, be louder than an ambient track.

As with music, listen critically. Do the SFX make the copy hard to understand? If there's *any* doubt, back the SFX off a bit, and listen again. Also, be sure to monitor your work through the studio speakers, not headphones.

If you're making use of punctuators, and assuming that your SFX are on a disc, the slip-cue technique (see Chapter 5) is absolutely indispensable. Using the slip-cue is the only way to guarantee the split-second timing necessary to make punctuators work.

Here's a hint to make voice tracks and noisy environments mix well:

Your voice should reflect how you'd speak if you were in the particular environment.

Here's an example. If you were talking with someone on the floor of a factory, chances are that, because of the noise around you, you'd make your voice considerably louder than it would in a more quiet setting. If you're setting your commercial in a factory, your voice should similarly show an increase in power. Note that I said to increase your *power*—not your *decibel level*. Set your mike level properly, and then back away two or three feet. Roll a tape, and speak loudly toward the mike. The Vu may still read considerably lower than it should. If so, crank up the pot until the level is okay, and then record. When you play the tape back, you'll see that even though the level is the same as it would normally be, the power in your voice is far greater. Voice power is very important, when you're using noisy ambient tracks.

When should you use SFX? When they make sense in the context of what you're doing. When they fit. When your work will benefit from their presence. As is the case with music, use of SFX should be governed first and foremost by thoughtful analysis of the copy and not by some knee-jerk reflex or whim. In general, if there's action in your copy, there should be SFX in your production.

Sound Effects Libraries

Copyright compliance isn't nearly the factor with SFX as it is with music. However, good SFX discs aren't likely to be available from your local record dealer. Again, search out reputable sound effects libraries. Here are a few:

The BBC Sound Effects Library—the best all-around collection available; tons of material (50 discs), well produced

The CBS Sound Effects Library—super ambient tracks, poor to fair on punctuators

The Electra Sound Effects Library—available on individual discs or as a set; great punctuators, fair ambient tracks; exceptional variety of popular sounds

The Valentino Sound Effects Library—great assortment of punctuators, pretty good ambient tracks; the quality of cuts isn't consistent

Finally, keep an eye out for things such as individual Electra discs, or Disney SFX discs. They sometimes show up (especially around Halloween) in record and/or discount department stores. The Disney discs, as you might suspect, are great, particularly the scary cuts (screams, creaking doors, howling dogs, etc.)

Though SFX libraries are not cheap, they are much more reasonably priced than music libraries. You can generally buy them outright (as opposed to leasing, as is the case with most of the good music libraries). Also, unlike music, SFX don't go out of fashion or become obsolete, as music nearly always does. Just ask anybody who bought a music library in the late 1970s, when libraries had a lot of disco cuts.

A final note about SFX libraries: at one time, there were libraries available on cassette tapes. I don't know whether they still exist, but if they do, forget them. Not only was the quality of the sounds watered down by tape hiss, but you could never cue up really tight to a cut. A cassette could never provide you with the split-second timing you'd often require, especially with punctuators.

Mixing Voice, Music, and Sound Effects

Blending all three elements into a single production is really no different from the voice/music and voice/SFX combinations previously discussed. I set the levels in their order of importance—first the voice track, again set slightly lower than normal to avoid overmodulation when the other tracks are added, then add any SFX; and then add the music. Once again, your primary objective should be to make absolutely sure that the voice track is 100% intelligible. Bear in mind that it's perfectly natural to feel like a klutz when you first try to manipulate three or more pots, but be assured that facility will come with practice.

Before you start mixing, make sure the head and tail of the recording are especially clean. If the music wows in at the start, or if your voice track begins with an audible click (you probably started the recorder to record your voice with the mike wide open), or if you hear *any* sound that shouldn't be there, *stop!* Check the individual tracks to locate the problem. If, for example, you have the notorious click on the voice track, cue the tape past the click. If the click is too tight to the voice, try cutting the click out with either a standard edit (see Chapter 10 on "Editing") or the invisible edit (see Chapter 11 on "Tricks of the Trade"). If none of these solutions works, recut the track. Don't let junk clutter up your opening. First impressions are of paramount importance here. If you begin a spot with garbage, you're going to have to spend the rest of the :30 or whatever trying to undo the error.

The same holds true for the tail of the spot. It's very tempting to just flip off the key switches to instantly end the spot. If you're confused, you don't have to worry about which pot controls which track—just flip them all off, right? *Wrong!* This practice makes your endings choppy. Always fade your pots out. Even if you fade your pots very quickly (in a half second or less), it'll still sound infinitely smoother than merely clipping them.

When you're fading many pots simultaneously, slide pots are decidedly more advantageous than rotary pots because with the slides, you can fade two or more channels with each hand, depending on your hand size. With rotaries, this would be virtually impossible. So if your board has rotary pots, and you're mixing sound from three or more channels, don't hesitate to use the Master output control which will fade *all* the channels together. Be sure to reset the master back to its appropriate level when you're finished.

At the head of the spot, if the music and the SFX begin together, wait :01–:02 before bringing the voice in. Don't let the voice start before the *aural* (i.e., hearing) stage has been set. The listener needs a second or two to get his or her bearings. After the voice is over, don't let the music and/or SFX trail out for more than :02. As has been mentioned, if the sound keeps trailing out, the listener isn't sure that the message is over.

Planning the Mix

Deciding how you're going to produce a piece of copy is a 100% cognitive process. There's no luck involved. Everything is planned and thought out ahead of time, preferably even before you go into the studio. That's not to say that hunches have no place in the creative process. They most certainly do, but first you must have a solid idea of what you want to accomplish. Hunches are spontaneous, but they often spring from ordered thinking.

Let's plan a mix. Here's a piece of copy, a station promo that I produced many years ago for WHEN in Syracuse, N.Y. The piece was intended for an 18–49, urban/suburban audience in a medium–large market (500,000 population). Also, because it was a station promo and not a commercial, I wasn't rigidly bound to :30 or :60 time limits. (Don't confuse freedom with license. Just because you might not have time restraints doesn't mean it's okay for your promo to be as long as *War and Peace*.)

```
 1  THE 62WHEN G.P. BASEBREAKERS TRAVEL TO MARCELLUS
 2  PARK TUESDAY, JULY 25 TO PLAY DONOVAN'S DILLONS
 3  FROM THE ONONDAGA HILL RECREATION LEAGUE IN A
 4  SOFTBALL GAME TO BENEFIT THE SYRACUSE JUVENILE
 5  DIABETES ASSOCIATION. THE ACTIVITIES BEGIN AT 6:45,
 6  AND JUST LISTEN TO THE PRIZES WE'LL BE BRINGING ALONG:
 7  A 40 CHANNEL C.B. RADIO FROM G.P. COMMUNICATIONS
 8  CASES OF BARRELHEAD ROOT BEER BARRELHEAD T-SHIRTS
 9  AND AN ESCAPE WEEKEND AT THE MARRIOTT, THE IN-CITY
10  RESORT. FROM 11am SATURDAY TO 3pm SUNDAY, YOU'LL
11  RECEIVE DINNER, BRUNCH, AND YOU'LL BE ABLE TO ENJOY
12  ALL THE MARRIOTT'S FACILITIES, INCLUDING THE POOL,
13  SAUNA, EXERCISE ROOM, AND GAME ROOM. AS THE
14  MARRIOTT CELEBRATES ITS FIRST BIRTHDAY, YOU CAN
15  CELEBRATE WITH THE ESCAPE WEEKEND THAT THE G.P.
16  BASEBREAKERS WILL BE GIVING AWAY AT MARCELLUS PARK
17  TUESDAY NIGHT.
```

There's nothing particularly outstanding about this copy. It's simple, straightforward, journeyman copy. It's up to the producer to make it sell.

The first thing to do is to look for breaks in the copy, places where the copy shifts focus, tone, or message. It's at these breaks that you can usually put punctuators, music changes, some sort of background activity. In this copy, two breaks seem fairly obvious: In Line 10, after the word "RESORT," the subject changes from softball game activities to the features of the hotel. In Line 14, after "GAME ROOM," the focus shifts briefly back to the game. Because the two subjects (game and hotel) were so different, I decided to highlight the difference by using two distinctly different pieces of music. Because the softball game had a connotation of activity, I made sure the opening piece of music had an uptempo, funky, driving feel to it. For the piece under the hotel section, I picked something completely opposite, a slow, lyrical piece with an almost semiclassical feel to it.

For the closing sentence, I had two choices. Either use another up-tempo composition, or reprise the opening piece. I decided on the latter course because bringing the first music back at the end would lend a cyclical, tie-the-ribbon-on-it sense.

Because the two pieces were so different, I couldn't very well just butt the two together following "THE IN-CITY RESORT." So I decided to smooth the juncture with a punctuator—in this case, silence. I abruptly dumped the up-tempo music in the middle of the phrase "AND AN ESCAPE WEEKEND AT THE MARRIOTT," and let the rest of the line continue with voice only, starting the mellow music after "IN-CITY RESORT."

That took care of the first music transition point. At the second, however, rather than use silence as a punctuator again, I opted simply to punch the fast music back in, following "EXERCISE ROOM AND GAME ROOM." Again, I consider running one piece of music directly into another to be rather risky. However, the combination of the obvious break in the copy and the solid punch-in of the music seemed to alleviate any problem. Also, I made sure that the reentry of the fast music continued the beat pattern of the slower piece, so the rhythm (for anyone who might be tapping a foot) would continue smoothly without a hitch.

I did one other thing with the music. When I selected the opening piece, I noticed that it contained a two-note pattern that repeated now and again. I also noted the series of prizes in the copy. Whenever I see lists of things I like to do something with them to make the items stand out. So I timed my reading so that the items on the list would fall neatly between and fairly tight to the punctuators. I think it worked pretty well.

As far as sound effects, I wasn't going to use any. I was content to let the music be the sole support of the voice. However, I decided that because the opening line talked about traveling to the park, I'd reinforce this with the sound of a sputtering Model T (modern cars don't sound funny or cute—not intentionally anyway) and a couple of suitable horn honks that I dropped in just prior to the entry of the music.

That's it (see Figure 9.5). You can hear the promo on the accompanying cassette. Let me say again that this is not an award winning piece of production. I wasn't trying to knock the audience's eyes (or ears) out. I wanted to take this copy and make it interesting enough so that we'd get a good crowd at the game and raise some money for the Diabetes Association. Happily, that's exactly what the promo accomplished. The planning took a couple of minutes, the production

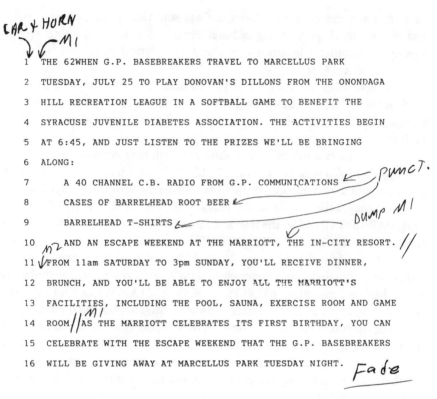

1 THE 62WHEN G.P. BASEBREAKERS TRAVEL TO MARCELLUS PARK

2 TUESDAY, JULY 25 TO PLAY DONOVAN'S DILLONS FROM THE ONONDAGA

3 HILL RECREATION LEAGUE IN A SOFTBALL GAME TO BENEFIT THE

4 SYRACUSE JUVENILE DIABETES ASSOCIATION. THE ACTIVITIES BEGIN

5 AT 6:45, AND JUST LISTEN TO THE PRIZES WE'LL BE BRINGING

6 ALONG:

7   A 40 CHANNEL C.B. RADIO FROM G.P. COMMUNICATIONS

8   CASES OF BARRELHEAD ROOT BEER

9   BARRELHEAD T-SHIRTS

10 AND AN ESCAPE WEEKEND AT THE MARRIOTT, THE IN-CITY RESORT.

11 FROM 11am SATURDAY TO 3pm SUNDAY, YOU'LL RECEIVE DINNER,

12 BRUNCH, AND YOU'LL BE ABLE TO ENJOY ALL THE MARRIOTT'S

13 FACILITIES, INCLUDING THE POOL, SAUNA, EXERCISE ROOM AND GAME

14 ROOM AS THE MARRIOTT CELEBRATES ITS FIRST BIRTHDAY, YOU CAN

15 CELEBRATE WITH THE ESCAPE WEEKEND THAT THE G.P. BASEBREAKERS

16 WILL BE GIVING AWAY AT MARCELLUS PARK TUESDAY NIGHT.

9.5 Copy marked and ready for production.

took less than an hour, and everybody was happy. On the down side, we were clobbered by the other team.

The point of all this is that except for the silent punctuator in Line 10, and the fortuitous existence of the two-note punctuator in the first piece of music, every bit of this production was planned and laid out before I got near the console. When I began the production, I already knew almost exactly how the promo would turn out.

Plan outside the studio; produce inside the studio.

Types of Projects

Before we leave the subject of mixing, you might as well familiarize yourself with the types of projects you'll be called upon to mix—namely, commercials (spots), public service announcements (PSAs), and station promos. Unless you're in the business of pure music production, you'll undoubtedly spend nearly all your production time turning out these items, so let's take a look at them.

Promos

I've just finished walking you through the production of a station promo, so you should have a pretty good idea of what the form is all about. It's basically a commercial for some aspect of your station. Promos can deal with new announcers coming on your air, contests for audience involvement, special programs or programming features, public appearances by your staff—anything that you want your audience to know about regarding your station. A promo's length can be whatever you want because you don't necessarily have to abide by the same time restraints as with spots. (Though most promos I've done approximate commercial length, I've also produced comprehensive station demos that ran at least 20 minutes. These were produced for national sales representatives who wanted to know everything about every aspect of the station's operation.) Also, assuming you've paid your ASCAP and BMI fees, you can use any music your heart desires. Finally, because you don't have a client looking over your shoulder, you're a lot freer to experiment with a promo. Don't get me wrong. Most clients are fine, and they're happy as long as you do your job well, but some feel they have to have a hand in every aspect of the job. How long a fuse do you have? If you're not sure, spend a couple of hours with the know-it-all client; you'll find out in a hurry.

At any rate, because of the lack (or lessening) of restrictions, promos provide you an arena in which to strut your stuff, show your mettle, grandstand for the folks—pick the cliché of your choice. However, just because anything *can* go into a promo doesn't mean that anything *should* go! Two elements to avoid at all costs are

1. poor (or vulgar) language
2. inside jokes

The Supreme Court notwithstanding, I don't know what passes for vulgar language in your community—or mine for that matter. I don't know at what point blue language becomes off-color, off-color becomes vulgar, vulgar becomes dirty, or dirty becomes obscene. Call it cowardice, but because I don't know where the boundaries are, I give the entire area a wide berth. Don't confuse freedom with license. If you police your own work and pass everything you do through a filter of common sense, you won't have to worry about somebody in your audience (or in your front office) doing it for you. You might call this self-censorship; I call it discretion.

Regarding inside jokes, have you ever attended a concert at which the performers giggled off mike to each other, or goofed around over things that they either didn't share with you or, if they did fill you in, you didn't know what they were talking about? If you want to make someone feel alienated, nothing will do the trick faster than putting him or her on the outside of an inside joke. Your station has probably worked hard to amass your audience, and radio listeners are, as a rule, a pretty loyal lot. However, if they don't understand what you're referring to, or if they feel you're talking down to them or around them to someone else, you'll cut your own throat. Do you know what a nanosecond is? In physics, it's 1 billionth of a second. In radio, it's the time it takes for an alienated listener to change stations.

"Won't a musician often put inside or personal allusions into a song?" I hear you ask. Sure, but let me ask you something. If you didn't understand those allusions, would you enjoy the song less? A lot of John Lennon's lyrics are obscure, autobiographical, or incomprehensible. Do you know who the walrus was (Goo goo ga-joob)? Does not knowing lessen the impact of the song? I don't think so. Why? Because for some reason, we're much more forgiving of artists sending us bizarre signals while toiling in their private creative fields than we are of radio voices coming across the public airwaves. Artists (especially the starving variety) might not care what the public thinks about their work; you don't have that luxury.

Avoidance of objectionable language and inside references is (or should be) germane not only to promos, but also to anything you produce. As is often the case, let common sense be the final arbiter. If you're unsure whether something is appropriate for your air, the chances are that it's not.

Commercials

Like them or not, commercials, as you've no doubt been told countless times, are the life blood of the mass media. According to tradition, Dr. Frank Conrad, who set up an experimental radio station (KDKA) in his East Pittsburgh garage in 1920, made a deal with a local record store owner. The merchant would provide free records to Dr. Conrad to play over the air in exchange for occasional plugs for the record store: thus, the commercial was born.

Good Spots and Bad Spots

As much as we profess to hate them, commercials are actually pretty interesting entities. They inform us about products and services; they mirror cultural and social change; and they can entertain. Our general aversion comes about because of two factors: their quality (too little) and their quantity (too many).

In general, when a TV program is really bad, when it treats us like dummies who need to be prompted by a laugh track, we generally (though not always) respond by not watching, and the source of the annoyance eventually disappears. Commercials are different. First, commercial producers aren't looking for a long run. A spot's lifespan is generally measured in days or weeks, sometimes in months, but almost never in years. Soon the spot disappears of its own will, having done or attempted to do its job. So though some spots seem to hang on forever like canker sores, most are born, live, and die within a fairly short time.

Another thing: producers of commercials want you to remember things. They seek retention. Don't unpleasant memories easily stick in your memory? Injuries, tragedies, embarrassments—you'd like to forget them, but you can't. As a result, especially on the local level where money is tightest, advertisers often strive to be stupid, figuring the more annoying the spot is, the more memorable it'll be. All too often, they're right.

In a world where the good guys always win, where justice always prevails, and where life is always good, bad commercials always fail. Unfortunately, you and I don't live in that world. We hang out in ours, and in our world, a commercial's success hinges not on its artistic merit, and not on its superior production, but on whether it sells a product. Whether it sells a product, in turn, hinges on whether the listener remembers the spot. To all too many advertisers, a negative impression is better than no impression.

To compound the injustice of it all, many fine spots have been wonderfully creative but have nonetheless failed to move the merchandise. The great Alka-Seltzer spots of the early and mid-1960s come to mind. There were three spots produced which were so popular that hook lines in the copy entered out popular language. Remember "Try it, you'll like it," and "I can't believe I ate the whole thing," and "Mama mía, that's a spicy meatball"? If you're too young to have seen these spots but still have heard these lines, recognize that their popularity is due to their use in those great commercials. However, although the commercials were wildly popular, and people often stopped what they were doing to watch them, and they were voted by the Hollywood Radio and Television Society as the outstanding commercial campaign of the 20-year period, 1960–1980, I've been told that they failed to sell Alka-Seltzer. People remembered the spots—but not the product.

All of which points to an ethical dilemma. If bad commercials often sell, and good commercials often don't, why bother to be creative? Why bother with quality? There are lots of reasons: pride, integrity, artistic growth. Do you realize that with audio commercial production, you're standing in the only arena in which art, science, and commerce share equal billing? Like a three-legged stool, if any single leg is too long or too short, you'll have an imbalance. If the spot's too artsy people will think you're being elitist and tune you out. If it's merely a showcase for production technology, it'll have all the artistic appeal of a computer printout. If it does nothing but hawk a product, you'll end up with one of those awful commercials that people write about in production books. What should you do? Take a cue from nature, and aim for a balance.

Commercial Formats

Here's a short (emphasis on *short*) overview of the types of commercials you're likely to come across. You're probably familiar with all of them, though you might not have realized it. There are many ways to organize types of spots. Here are the headings I use:

- problem-solving
- slice-of-life
- narrative
- demonstration
- spokesperson and testimonial
- musical

The Problem-Solving Commercial

Though most spots tell a story of some kind, this form deals with a specific problem, and a specific feature of the product that will solve the problem. Wisk takes care of ring-around-the-collar. Bounty towels pick up spills quickly. Often these spots are addressed directly to the listener. "Do you have bad breath?" "Are you troubled by dandruff?" "Is underarm odor a problem?" (Obviously, this person hasn't had a date in a while!)

The Slice-of-Life Commercial

More general than the problem-solving form, this type of commercial often focuses on the whole product rather than a specific feature. Here, as in a play, we're likely to find a beginning, a middle, and an end. Unlike the previous form, we're observers of the action and not the focus of it. We watch the Maytag repairman while away the hours. We watch the man or woman arguing with a tub of Parkay margarine. We watch as a cordon of lady shoppers confront Mr. Whipple while he's squeezing the Charmin. (Mr. Whipple hasn't had a date in a while either.)

The Narrative Commercial

The narrative spot, by definition, involves someone talking directly or indirectly to you, rather than you watching something transpire. The narrator (and/or the action) creates a situation or environment that's personal, involving, and frequently emotional. A bunch of guys sit around the campfire with their bottles of Lowenbraü toasting "Here's to good times and good friends." Mean Joe Greene trades his overripe football jersey for the little boy's Coke in one of the most honored commercials of all time. It's moving day, and as the movers load the truck, Mother sits in the attic amid the cartons, lovingly leafing through the box of Hallmark cards that chronicles her life. The retired gentleman, tentative, and obviously apprehensive, sets out for his first day working at McDonald's, and by the end of the day is such a success, he tells his wife "I don't know how they got along without me." It's a great spot. Truly wonderful (although it's a rather sad commentary that after a life of hard work, this man has to flip burgers to make ends meet). In terms of pure memorability, it's hard to beat a good narrative spot.

The Demonstration Commercial

Ever watch those people late at night on the cable stations who sell all sorts of amazing gadgets? Slicers, dicers, shredders, whippers, miracle car waxes, miracle ironers, miracle stain removers, miracle sharpeners, etc. I know that these program-length commercials bug a lot of people, and the claims they make may not always be valid, but these shows are still fun to watch. Why? Because the eye (and the ear, remember?) likes movement. There's something oddly fascinating about watching a machine spew out mounds of julienne fries (whatever they are!) in just seconds. Here, the demonstration takes the place of explanation.

Salespersons since the dawn of time have known that nothing impresses a prospective buyer as well as a demonstration. That's why the snake-oil pitchman would call an old man (capable of only limited movement and negligible agility) up to the stage, administer a dose of Dr. Elmo's Mystery Elixir, and watch in self-satisfied amazement as the fellow began turning cartwheels and back handsprings.

Don't make the mistake that many do and assume that because radio isn't a visual medium, the demonstration spot won't fly. Remember that radio *does* deal with visual images—the ones inside the listener's mind. Use your sound effects library to create mental pictures. I may not be able to see the julienne fries pouring out of the Acme Slicer/Dicer, but I can hear the machine working away, and I can hear the audience "oo" and "ah" over what's happening.

The Spokesperson and the Testimonial

Marketers have long known the value of a celebrity spokesperson. What's it worth to have Bill Cosby smile and say nice things about your product? A lot. Just ask the teams at Kodak and Jell-o and E.F. Hutton. Athletes, movie stars, politicians—anyone who's famous at the moment—can lend his or her name, face, or voice to a product, and chances are the product's sales will benefit. Wilt Chamberlain showed how he could fit into a Volkswagen. Lauren Bacall purred over Maxwell House coffee. Tip O'Neill popped out of a suitcase in a motel room. Ed McMahon is one of the best around today. I once heard him say "If I can hold it up or point to it, I can sell it." However, notice that although a celebrity may show and laud a product, the celebrity may not actually say "I use it." Now we're in the realm of the testimonial.

With a testimonial, the person speaking for the product actually uses it. When Lee Iacocca pitches Chrysler cars, you can bet that he drives one. On the radio, Paul Harvey professes never to advertise a product that he himself doesn't use. Unlike the spokesperson who, by definition should be well known if the association is to rub off onto the product, the testimonial can feature anyone. Generally, a testimonial will feature an average person (whatever that means) or an actor playing the part of an average person with whom the general audience can identify or at least sympathize. Isn't it interesting? We buy things because rich-and-famous people tell us to, and we buy things because ordinary people like us tell us to. Are we pushovers or what?

The Musical Commercial

I'm not going to spend much time here because we'll cover the subject more thoroughly in Chapter 14, on "Jingles." Suffice it to say that music is a powerful sales tool. In New York State, the tourism department is still using the same four-note "I Love New York" jingle that it began using back in the late 1970s. Why? Because it still works. However, just because you don't have a big studio, don't think you can't do musical spots. Run through your music library, find a piece of music you like, write lyrics that fit the music and your purpose, and produce. I produced (and still produce) a lot of promo songs for stations, organizations, and causes I support just by laying original words over an existing music bed.

Comparing and Contrasting the Commercial Formats

If you think that there's an overlap among these various commercial styles, you're right. Many terrific spots combine elements from two or more styles. Don't pigeonhole yourself by attempting to write commercials by a formula. Remember that you want to adapt the style(s) to the product, not the other way around.

Let's say you're doing a commercial for a used car dealership. In a problem-solving format, you might address the families in the audience on a tight budget, showing how the dealer can find them a great car within their means. In a slice-of-life format, you might show us a couple as they find a car they can afford from among dozens of models on the lot. A narrative approach might focus on the friendly (Aren't they always?) salespeople who care for their customers and take the guesswork and apprehension out of buying a car. Want to use the testimonial approach? Hang out at the lot, interview a few satisfied customers, and write the copy around their remarks. Need a spokesperson? You don't have to go to a nationally recognized individual. A local celeb will do. Get hold of the mayor, the local football coach, a star athlete, the head of the Chamber of Commerce, a prominent attorney, someone well known for charitable contributions—any of these people can be of use.

The Humorous Commercial

Actually, the humorous spot isn't so much a commercial format as much as it is a style. Nevertheless, I didn't want to leave the subject of commercials without touching on humor because humor is an area where many consider themselves adept, and few really are.

Some of the funniest announcers I know aren't particularly funny off-mike, and some of the funniest people I know die when the mike opens. Go figure. Humor is a funny thing (sorry!). It's astonishingly nebulous; witness the legions of comedians currently plying their trade. I can't define humor. My dictionary defines it as, among other things, "that quality which appeals to a sense of the ludicrous or absurdly incongruous." No wonder Webster never played Vegas! Laughter itself is funny. Why do we do it? What causes it? Why does our body respond with this odd reaction to silly stimuli? I haven't a clue. I don't know why things are funny. All I can do is to examine the things that are. If I can find a common thread, great. If not, well at least I had fun.

Nothing is less funny than an unfunny person trying to be funny. Remember Robin Williams's immediate superior in "Good Morning, Vietnam," the humorless little Napoleon who said "I know funny." This guy got on the air and proceeded to be about as funny as an abscess. Really funny people are extremely rare, and comedy doesn't age too well. Most practitioners come and go in a flash—which makes people like George Carlin, Robert Klein, Robin Williams, Steve Martin, and Jay Leno, who've stood and will continue to stand the test of time, so remarkable. Novice announcers *all* try to be funny. It's amazing. For some reason, they believe that the audience expects lots of laughs, and, by George, what the audience wants, the audience gets. Yucks by the pound. Help! These beginners fail because they're trying to be something they're not, and because they're trying to find funny things rather than understanding what things are funny.

I break humor into two general categories:

1. word humor
2. situational humor

Word Humor

Word humor draws its roots from the idiosyncracies, oddities, sounds, and meanings of our words. Here are a few of my favorite forms:

- pun
- satire and lampoon
- understatement
- overstatement
- malapropism
- anachronism

Pun

Though at the top of the groan scale, a good pun is golden. A *pun* (as everyone who had to sit through Shakespeare knows) is a play on words. You need a word with two meanings. You use the word in a situation where one of its meanings is implied, but instead you switch to the other. If you want puns, take a graduate course in Groucho. "I've got a good mind to join a club and beat you over the head with it." Abbott and Costello's immortal routine "Who's on first" is nothing but a flood of puns, with *who, what, I don't know, today, tomorrow, why,* and *I don't give a damn* having not only their usual meanings, but also being the names of the baseball players.

Satire and Lampoon

These forms of word play poke fun at people, organizations, anything you can get in your sights. *Satire* is generally more subtle; *lampoon* is broad, like burlesque or farce. Again, look to Groucho, who could unstuff a shirt better than anyone. His exchanges with the wonderfully stuffy Margaret Dumont are priceless.

GROUCHO: Not that I care, but where is your husband?
MARGARET: Why, he's dead.
GROUCHO: I'll bet he's just using that as an excuse.

GROUCHO: I love you. Why don't you marry me?
MARGARET: Why, marry you?
GROUCHO: Married! I can see you right now in the kitchen bending over a
    hot stove. . . . But I can't see the stove.

Stan Freberg, George Carlin (who really loves playing with words), Robin Williams, Fred Allen, Ernie Kovacs—we've been blessed with a wealth of great satirical minds. Some talented people couple words and music for satirical effect. If you've never heard (or heard of) Allan Sherman ("Hello Muddah, Hello Faddah"), Spike Jones, Tom Lehrer, or Victor Borge, get to the nearest library or used record store, and listen.

Understatement and Overstatement

Understatement involves downplaying reality. The Black Knight in "Monty Python and the Holy Grail" has an arm lopped off in a duel and exclaims "It's only a flesh wound." Columnist Dorothy Parker once assessed Katherine Hepburn's acting ability as "running the gamut from A to B." The converse, overstatement, involves stretching and exaggerating reality. Humorist Wilson Mizner, expressing displeasure to the waiter in a fancy restaurant remarked, "I've had better steaks than this for bad behavior." Mizner was also the man who likened Hollywood to "a trip through a sewer in a glass-bottomed boat."

Malapropism

This is based on Mrs. Malaprop, a character in Richard Sheridan's 1775 play *The Rivals*, who was known for her unintentional misuse of language. For example, Mrs. Malaprop wanted her daughter to study geometry "so that she might know something of contagious countries." The two greatest practitioners of the form in my lifetime have to be baseball great Yogi Bera ("If people don't want to come out to the park, nobody's gonna stop them"), and movie mogul Samuel Goldwyn (When told by an underling that a script was too controversial because it was about lesbians, Goldwyn supposedly replied, "All right, where they got lesbians, we'll use Austrians"). Goldwyn was the man responsible for gems such as "A verbal agreement isn't worth the paper it's written on," and "Anyone who goes to a psychiatrist should have his head examined." Oh yes, let's not forget the inimitable Casey Stengel who, in his old age, when asked how he was feeling, replied, "Not bad. Most people my age are dead."

Anachronism

When something (a word, a phrase, an object, a well-known character) is chronologically misplaced, particularly when its misplacement is obvious, it is an anachronism. Shakespeare has a few of these to his credit. In *Antony and Cleopatra*, Cleopatra plays billiards, a game that didn't exist in ancient Egypt. In *Julius Caesar*, Cassius tells Brutus that "The clock hath stricken three," when, in fact, there were not striking clocks in Caesar's time. The television series "Moonlighting" won a number of Emmy Awards for a hysterical retelling of *Taming of the Shrew*, in which Petruchio's horse not only has a blanket with a BMW logo, but also wears sunglasses. Petruchio himself enters with a hearty "What's shakin', y'all?" At his wedding to the shrewish Kate ("Funny, you don't look shrewish!"), the minstrels in attendance segue from 'Greensleeves' into "Good Lovin'." This is anachronism at its best, and it's worth catching this episode on the rerun if you can.

Situational Humor

The area of humor draws not from language, but from the interplay of characters and situations.

Dick Orkin and Bert Berdis, the funniest radio and advertising duo of the 1970s, when asked where their ideas came from, replied "silly people in ordinary situations and ordinary people in silly situations." It's the conflict that's funny. That's why comedy teams always have a straight man and a gag man. Think about it. Eddie Murphy's and Robin Williams's movies depict anarchy in the midst of conservatism. Peter Sellers's Inspector Clouseau was a naive boob waging war on brilliant criminals. Monty Python alumnus John Cleese went both ways. He was either a stuffy, staid Englishman surrounded by madness (the "Parrot Sketch," the "Cheese Shop Sketch"), or a slightly off-kilter fellow seeking to function in a mundane environment (the "Minister of Silly Walks").

A naive person in the worldly setting is a virtual staple of comedy, especially television comedy: Edith Bunker on "All in the Family," Woody Boyd on "Cheers," Radar O'Reilly on "MASH," Gracie Allen, Mary Tyler Moore, Bob Newhart, Jackie Gleason's Poor Soul, Art Carney's Ed Norton, Tom Smothers, all of the Beverly Hillbillies, Goldie Hawn—the list is pretty long. Also, look at the great silent clowns: Chaplin, Keaton, Langdon, Lloyd, all struggling to hang onto their dignity in a world with nothing better to do than take it from them. The great ones all knew that the boundary between comedy and tragedy is extremely fine. They also knew that there's a difference between stupidity and naivete. Both generate laughs, but only naivete can generate sympathy.

Enough! If you're into humor, get to the library and your local record store and do what will probably be the most enjoyable research imaginable.

PSAs

Since the deregulation of the radio industry, public service announcements (PSAs) aren't nearly as common as they once were. True, program directors would often shovel the PSAs into the wee hours of the morning (especially Sunday) where no sane advertiser would want a commercial to be. The station would fulfill its public service requirement to the FCC (Federal Communications Commission), and the kid working the overnight shift (for minimum wage most likely) would have something to play (besides records) to help him stay awake.

A PSA is like a commercial except that the client isn't trying to sell a product for a monetary profit—though for an emotional or spiritual profit perhaps. However, aside from the funds needed to keep operating, they're not out for your money. The Red Cross, the Boy/Girl Scouts, organizations fighting specific diseases (Cancer Society, Leukemia Society, Lupus Foundation, Diabetes Association, etc.), organizations promoting healthy body parts (Heart Association,

Lung Association, Kidney Foundation): The list of worthy (most of them, anyway) organizations that run PSAs on radio and television is a long one. Like promos, PSAs don't have to fit the time strictures of a commercial, but most of the PSAs you're likely to run across, especially if they've been done for large, nationally known organizations, will probably be produced in :30 or :60 lengths.

Don't confuse public *service* with public *affairs. Public affairs* deals with two-sided (or more) issues: abortion, gun control, minority rights, prayer in schools, environmental maintenance—anything controversial falls under the heading of public affairs. Politicians making an editorial comment on your air may think they're performing a public service, but chances are they're merely serving their own agendas, regardless of how universally held their opinions may be. On your program log, label their spots PA and not PS.

PSAs can use all the approaches a spot can. Sally Struthers asks you to help save starving children as she walks through a Central American slum. Yul Brynner asks you to stop smoking in a PSA produced just prior to his death from lung cancer. Two talking crash dummies want you to buckle your seat belt. The Mormons (Church of Jesus Christ of Latter-Day Saints), who always produce good, emotion-driven, ecumenical PSAs, want you to get along with yourself and others better. An egg plops onto a hot skillet and you hear "This is your brain on drugs." Any form works with a PSA. Most good ones, though, have a strong undercurrent of emotion throughout—which makes sense if you think about it. With a PSA, you're asking the listener to care about something, and the way to a person's heart is through the emotions, not the intellect.

One of the most powerful PSAs I remember was done by the Red Cross many years ago. As I recall, a man was sprawled out at the bottom of a flight of steps, his body being half on the lower steps, and half on the sidewalk. People, who were shown only from the neck down, were walking around him, stepping over him, generally ignoring his condition. The message was that if you are able to donate blood but don't, you're no better than the people refusing to acknowledge the plight of the man on the ground. The PSA didn't say this outright, of course, but the implication was there—and it was strong—and successful. I hate giving blood, but so far I've been tapped for three gallons—all because of that PSA.

Summary

One final word. It's awfully hard to learn in a vacuum. If you're relying solely on ideas from your own mind for inspiration, you probably won't get very far. Seek out and listen to what's good. The Radio Advertising Bureau (RAB) in New York has cassettes of every year's Clio Award Winners. Prize-winning PSAs and promos are also available from a number of sources, like the Hollywood Radio and Television Society. Listen and learn. There's a big

difference between learning from someone else's ideas and methods, and pla-
giarism; between using someone else's work as a source of inspiration, and
stealing that work. Every time you learn something new, your arsenal of production ideas grows geometrically; a single idea combined with things you
already know makes for a wealth of new possibilities.

Review

1. Much of radio production involves combining voice, music, and sound
   effects (SFX).
2. Before recording, align your Vu meters, and clean the equipment.
3. Foreground/background interaction occurs when there is a momen-
   tary pause by the voice (foreground) and a concomitant slight eleva-
   tion of level in the background (music, SFX). By shifting the focus,
   even for a short time, we hold the listener's attention.
4. In most cases, produce the voice track first. Don't go on to any other
   phase of the production until the voice track is perfect.
5. Your choice of background music should be determined by the needs
   of the commercial. The tone of the music should reflect the quality of
   the voice and the nature of the product. Never use music containing
   lyrics that are sung.
6. Listen for musical punctuators, and use them if at all possible, even if
   it means adjusting the copy slightly. Instant silence makes a good
   punctuator.
7. Use multiple music tracks if the spot clearly changes focus or tone.
   Don't butt two pieces of music together; join them with an edit (keeping the rhythmic flow), or separate them with a short pause. The
   music tracks should be completely different in tone, tempo, and instrumentation.
8. A music track can end cold (on a single note) or can fade out.
9. When mixing voice and music, start with the voice at a solid level,
   and gradually increase the music level. When the music starts to
   interfere with the voice, back off a bit.
10. Monitor your mix through the studio monitors and not through the
    headphones.
11. Watch the clock. Stay within your time constraints.
12. Music libraries are the best source of good production music.
13. Watch out for copyrighted music.
14. There are two classes of sound effects: punctuators and ambient
    sounds.

15. Speaking over an ambient track, your voice should reflect how you'd speak if you were actually in that environment—that is, with increased or decreased power, depending on the environment.

16. When mixing voice, music, and SFX, set the levels in order of importance—voice level first, then the SFX, and finally the music.

17. At the end of the mix, fade out, rapidly, if you like. Don't merely flip the key switches off; you'll risk putting a noise on the tape. If you have to fade a number of rotary pots simultaneously, use the master gain control.

18. Plan your mix before you get to the studio.

19. The three most common production projects are station promos, commercials, and public service announcements (PSAs).

20. A promo attempts to sell some aspect of your station to your audience. Promos don't have the same music and time restraints that commercials do.

21. Promos give you the opportunity to experiment and to be creative.

22. Avoid vulgar language, as well as references your audience won't understand.

23. Commercials seek to be remembered. Often, poorly produced spots, because they're so bad, stick in the memory, so producers are sometimes tempted to deliberately turn out inferior work.

24. Aim for a balance between the artistic, the technical, and the commercial aspects of the spot.

25. Some of the commercial formats include problem solving, slice-of-life, narrative, demonstration, spokesperson, testimonial, and musical.

26. Many good spots combine elements from two or more formats.

27. Almost any format can be adapted to promote almost every product.

28. Many people try to be humorous, but few succeed.

29. Humor can be classified as either word humor or situational humor.

30. Word humor forms include puns, satire and lampoon, understatement, overstatement, malapropism, and anachronism.

31. Situational humor involves either silly characters in mundane situations, or mundane characters in silly situations.

32. A PSA is like a commercial except that the advertiser isn't trying to sell a product for monetary profit.

33. PSAs don't have to abide by the same time constraints as commercials, but most nationally produced PSAs are produced in standard commercial lengths.

34. PSAs deal with noncontroversial subjects. Announcements concerning controversial subjects should be classified as *Public Affairs*.

10 Editing

Few aspects of production can be more rewarding or can give you a greater sense of accomplishment than editing. On the other hand, you need to have a really strong sense of self-worth, because if you're good at editing, you'll rarely receive any compliments for it. Good edits go unnoticed. Having people tell you how much they like your editing is like having them tell you how natural your toupee looks.

Aside from recording, editing is the single most important skill the production professional must master. Good editors can not only remove mistakes from a track and smoothly conjoin tracks recorded at different times in different places, they can also create novel effects. George Martin, the Beatles' producer likes to tell how he created the weird calliope sound for the *Sgt. Pepper* album. He recorded a calliope track onto tape and told his assistant to cut the tape into foot-long pieces. When that was done, Martin took the pieces, threw them into the air, and told his befuddled assistant to splice the pieces together in random order. That spaced-out calliope sound on the album is the result.

Although editing a voice track is not as complicated as editing music, neither is particularly difficult, which is good news because your edits *must be perfect*. If people tell you that nothing's perfect, don't believe them. If your edits are less than perfect, all the work that's gone into your recording is wasted.

Basic Tools (see Figure 10.1)

- a splicing block
- a single-edge razor blade
- splicing tape
- grease markers

1. *Splicing Block*. The standard audio block has a ¼-inch-wide horizontal groove in which the tape is positioned during editing. This groove is intersected by two narrow guides for the cutting blade, one of which guides cuts through the tape groove at a 45-degree angle, the other at 90 degrees.

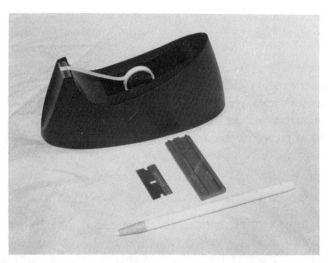

10.1 Splicing tools.

Some tape decks have built-in splicing blocks, frequently mounted on the head cover. There are cheap blocks available at your local electronics hobby shop, but I'd stick with a well-known, professional quality block, such as the Edit-Tall block, named for Joel Tall, the fellow who designed it in the early 1950s. The Edit-Tall block is solid, stable, easy to use, and can be bolted down to the table for additional stability. I like to have a separate block at each tape deck.

Some blocks have an additional tape groove, ⅛-inch wide for editing cassette tape. I try to stay as far away from cassette editing as I possibly can. The tape's too thin, and the editing process too imprecise to be anything but an annoyance.

2. *Single-edge razor blade.* Check your local pharmacy. Like most things, these blades are cheaper if you buy them in bulk. Do so. You want to be able to grab a new blade the instant the old one seems dull.

3. *Splicing tape.* You'll only find this available at electronics supply stores. Unlike everyday transparent tape, splicing tape is thinner and has much less adhesive. *Never* use household transparent tape for editing. In fact, never use anything but splicing tape. This is no mere professional bias. That roll of tape from the kitchen utility drawer has some serious drawbacks. First, it's more than ¼-inch wide, so you'd have to trim it to fit the audiotape. Second, the excessive adhesive can (and will) ooze out under pressure and will gunk up adjacent layers of tape. Third, the excessive adhesive can befoul tape heads, capstans, pinch rollers, tape guides—anything it comes in contact with. Some companies manufacture precut lengths of splicing tape. I've never used them,

but the idea of someone else cutting my splicing tape for me seems a trifle extravagant—sort of like having someone butter your bread for you.

4. *Grease markers.* I'm partial to Berol White China® markers. They're available at any good stationery or office supply store. Also, department stores or drug stores with stationery departments should have boxes of Berol markers. The brand of marker you use isn't particularly critical, but the color is. Always use white. Most recording tape today is brown or black, and a white mark is easier to see against these backgrounds than any other color.

Voice Editing

In its simplest form, this is the procedure for editing something from a voice track. Let's say we record you saying "I don't like broccoli," and in a fit of whimsy we decide to edit out the word *don't*. Here's what we do:

1. Rewind the tape to the head of the recording.

2. With the tape up against the heads (in cue mode), manually move the tape ahead until you hear the *d* in the word *don't*. Because you're moving the tape very slowly, the voice will, of course, sound muddy. Don't worry. If you listen carefully, you'll be able to distinguish individual sounds, especially hard consonants like *d*. Cue the tape to the very beginning of the *d*. This sound is now positioned at the center of the playback head (assuming your deck is set to playback or to repro. Remember, if you're set on sync or on sel-rep, the tape is playing back at the record head, and you should make your grease marks there.). Make a *small* mark (such as a thin vertical line) on the back of the tape (not on the tape's recording surface that's against the head) at the center of the playback head. Make the grease mark just large enough to be seen, and make sure you don't mark the head itself. Grease on the playback head can wreak havoc with your sound reproduction.

3. Move the tape ahead until you hear the *t* sound in the word *don't*. Then continue until you hear the start of the *l* in the word *like,* and make your second mark just before the *l*. "Wait a second," I hear you exclaim. "Why not make the second mark immediately after the *t?* Simple. Remember that our goal is not merely to cut the word *don't* from the voice track, but also to do so *while maintaining the natural flow of the language.* If you analyze "I don't like broccoli," you'll note there are minuscule pauses between the words. If we simply remove the word *don't* from the recording, we'll end up with *two* pauses between *I* and *like,* and the voice won't sound natural. By making our second mark at the head of *like,* we will cut out not only the word *don't,* but also the pause between

don't and *like*, thus keeping the cadence of the language smooth, as shown in the accompanying figure, illustrating where the editing should take place. The figure is written in reverse since, if you think about it, that is the direction the words are imprinted on the tape.

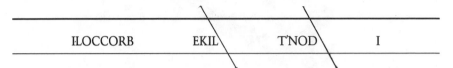

4. Manually loosen the tension on the tape, so that you can lay the tape into the groove on the splicing block.

5. Position the tape so that your first mark is bisected by the 45-degree blade guide (see Figure 10.2).

6. Cut the tape at the 45-degree angle (see Figure 10.3).

7. Move the tape over so that your second mark is similarly positioned in the 45-degree slot, and cut through the second mark. You've now edited the word *don't* as well as the space between *don't* and *like* from your tape. You might be tempted to throw this little piece of tape away, but hang onto it until you've heard the final product. If your edit turns out to be less than perfect, you can always reinsert the edited piece, and start the process again. However, once you've tossed the piece into a wastebasket full of bits and pieces of tape, retrieval is virtually impossible.

8. Butt the two tape ends together in the tape groove (see Figure 10.4). There should be absolutely no space between them, nor should there be any over-lapping. The tape should also be positioned so that the seam doesn't lie over either of the blade guides.

9. Cut off a ½-inch to ¾-inch piece of splicing tape, lay it over the seam (see Figure 10.5), and press down. Gently use your fingernail to work any air bubbles out from under the splice.

10.2 Placing the tape in the splice block.

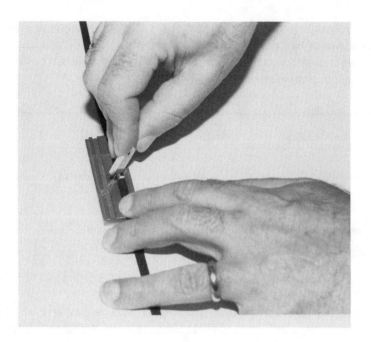

10.3 Cutting the tape.

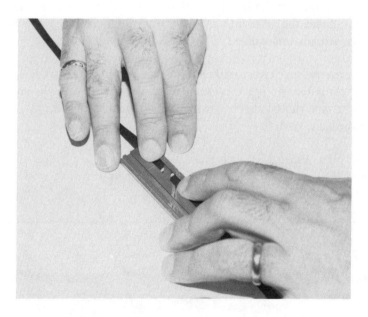

10.4 Butting the tape ends together.

10. Carefully lift the tape from the block, keeping hand contact with the oxide (recording) side of the tape to a minimum.

11. Rethread the tape, and listen to your edit. If you've done what you intended, you should hear yourself exclaiming, "I like broccoli."

Miscellaneous Notes on Editing

Several tips may help you perfect your editing and may make editing easier for you.

Not Touching the Tape

If you're careful, you can go without touching the oxide side of the tape until Step 11 of the editing process. If you have to touch the oxide side, do so only on the piece you're cutting out. You can hold the tape on its edge when placing it in the block.

Demagnetizing the Blade

Before editing, demagnetize your blade. The bulk eraser you use to erase reels, carts, and cassettes can also be used to remove any magnetism

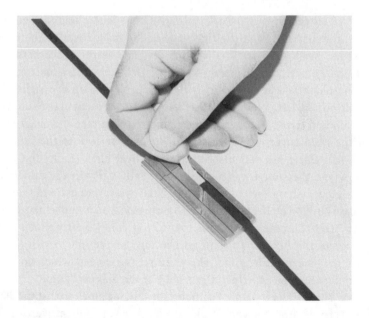

10.5 Taping together the splice.

that may have built up on the blade. If the blade is at all magnetic, the magnetic field could affect the signal on the tape at the points you're cutting.

Choosing the Best Angle
for Cutting

In general, use the 45-degree blade guide to cut the tape. It's longer than the 90-degree cut, so the splice will be stronger. Also, because of the angle, the force pulling across the splice is less than it would be pulling at exactly a 90-degree angle. In addition, if you look carefully at a 45-degree angle splice, you'll see that there's a tiny bit of overlap, a small stretch (¼-inch or less) where tape on both sides of the splice plays simultaneously. As a result, this overlap makes the transition from the right side of the splice to the left smoother, especially with music, as is shown later in this chapter. In case you're wondering, the 90-degree cut is good for editing between words or syllables that are so close together, the diagonal cut won't fit. In addition, the 90-degree cutting slot on the splice block can be a reference for editing without a grease pencil, or at the very least, being able to edit without marking the playback head, as described in the next tip.

Using a Reference Point
for Cutting

It seems that on a lot of older tape decks, such as Scullys and Ampexes, the distance from the right side of the head cover to the center of the playback head was exactly the same distance as the space between the vertical (90-degrees) and diagonal (45-degrees) cuts on the standard splicing block. The block was mounted with the 90-degree cut on the right, the 45-degree cut on the left. When the tape was cued to a particular point, instead of being marked at the head where the sound was, the tape was marked at the edge of the head cover, a short distance (an inch or two) to the right of the sound. When this mark was placed in the vertical blade guide, the point on the tape where the sound was located would fall neatly in the middle of the 45-degree guide, where the cut was made. Thus, you were assured that there'd be no grease buildup on the heads. If, for some reason, you found yourself without a marker, after the cuing point was found, you could lightly pinch the tape at the edge of the head cover, and set the tape in the block with the pinch mark in the vertical blade guide. The point on the tape at which you wanted to cut would, again, be exactly in the middle of the diagonal guide.

This is a handy fact to know. Position your splicing block so that the diagonal cut is directly in front of your deck's playback head, and the vertical cut is off to the right. Does the vertical cut line up with anything you could use as a reference? Tape guide? Head cover? A screwhead? Anything will work. If there's nothing you can use as a reference, there's nothing wrong, aesthetics notwithstanding, with cutting or drawing a reference mark on the tape deck itself. I know that a lot of people will object to marking the equipment in any way. However, if you do a lot of editing, I see no problem with making a small, permanent reference mark on the deck. If you're heavy-handed, marking the tape at the heads can not only result in grease build-up on the heads, but also affect the heads' alignment. If marking the gear makes you squeamish, but you still see the value of a reference mark, make a temporary mark on the deck, using your grease pencil, and when you're done, clean off the mark with alcohol.

Making Big Cuts

Our "I don't like broccoli" edit was made all the easier by the fact that the two edit marks were so close together. However, it's not unusual for the edit marks to be many feet apart. Let's say you're editing a 10-minute taped speech and decide to lop off an entire paragraph. Obviously, the space between the marks, between the beginning of the outgoing paragraph and the start of the following paragraph, could be considerable. This is where the *edit* feature of your tape recorder is invaluable.

Make your two marks, and rewind. Make your first cut, at the head of the paragraph you're cutting. Then, instead of spooling off the tape from the supply reel in a frustrating search for your second mark, thread the tape from the supply reel through the capstan and pinch roller, and then push *edit* (on some decks you'll push *edit* and *play*). Even though the tape isn't wound onto the takeup reel, it will be pulled along by the capstan and pinch roller, and will slowly spill onto the floor (or into a wastebasket, if you're gutsy). In addition, the tape will be *playing*, so you can hear when your second mark is about to come along. It beats eyeballing a mile of tape for a grease mark.

Preparing for Editing
While Recording

If you're doing a :30 spot, and you trip over a word, you have two choices:

1. Start Over.
2. Pick up from the mistake, continue to the end, and then edit the mistake out.

This is a judgment call. If you goof within the first few seconds, do another take. Don't waste time with an edit. In fact, unless the read has been really good, even if you trip after :20, start from the beginning again. Obviously, though, if you're 15 minutes into a 20-minute audio track for a slide presentation, you're not about to take it from the top. If you do trip, do this:

1. Stop!
2. Let the tape continue rolling for about :01–:02.
3. Say "Take 2."
4. Resume your read, either with the word you tripped over, or from the beginning of the sentence. Picking your read up in the middle of a sentence is tricky because the voice cadence you use from the middle of the sentence to the end has to match the cadence you used in the beginning. It's easier and less likely to sound funny if you take the entire sentence from the top.
5. If you trip more than once over the same word or phrase, you can let the tape roll, and slate each take (Take 2, Take 3, etc.), making sure you keep a log of where your trips are, and how many takes you took at each error.

Deciding Where to Make the Cuts

When you move the reels manually, positioning an edit point at the playback head, you'll find that some sounds are easier to cue to than others. Consonant sounds like *b, d, g* (as in *get*), *j, k, p, q, t,* and *x* are easy to find because they're somewhat explosive. Also, prior to pronouncing them, we're momentarily silent as we build up the air pressure necessary to sound them. Because of this minute silence preceding these consonant sounds, we can easily mark where they begin.

The same cannot be said of consonant sounds like *h, l, m, n, r, w,* and *y*. Record the word *hammer*, and try to pinpoint exactly where the *h* or the *m* or the *r* begin. It's not easy because these sounds and the adjacent sounds seem to flow together, with no distinct points of demarcation. The same is true of most vowel sounds. As a result, for ease and accuracy, you should always try to use explosive consonants as your edit points.

If, for example, I say "The boy was hiking in the woods," and I trip on the word *hiking*, I could opt to pick up from "hiking" and continue. When I went back to edit, rather than make my marks at the beginning of each "*hiking*," I'd rather split the words, and mark at the *k* sounds because the *k* sounds are so much easier to cue to than the *h*'s. The edit would look like the accompanying figure.

..... SDOOW   EHT   NI   GNI\KIH ....'NI\KIH   SAW   YOB   EHT

Of course, it would be best to edit from the beginning of the sentence, but should that be impossible for some reason, and should you have to edit at the word *hiking*, you're much better off editing to the *k* than the *h*. Don't let the fact that you're editing in the middle of a word scare you. If your marks are placed correctly, you have nothing to worry about. When you've made your marks, you can double-check to see whether they're correct by doing the following:

1. Cue your first mark to the center of the playback head.
2. Play the tape, and listen to see whether that point is where you want your edit. If it is, fine. If not, make another mark, slightly to the left or right of your original mark, depending on whether that original mark was ahead or behind the edit point you want, and check again.
3. Repeat the process with your second (original) mark.

Editing in Midsentence

If you have to pick up a read in midsentence, or even in midparagraph, it's a good idea to roll the tape back a bit and listen to the playback to make sure you're continuing the same pace and cadence. The speech patterns on both sides of the edit *must* match. If anything's different—pitch, pace, dynamics, cadence—the listeners will be able to zero in on your splice like bees on honey. They may not say to themselves, "Gee, what a lousy splice," but they'll certainly know that something is not right. They'll also be able to tell you exactly when it happened. Never underestimate the ears of your audience.

Music Editing

If you've had any form of music background, editing music will be a breeze, but even if you haven't, don't worry. Anybody who can count to four or tap a toe to music can learn the basic techniques.

Rhythmic Continuity

Note: A splice must *never* cause a break in the rhythm.

Most commercial music you'll run across will follow the conventional rules of Occidental music. If this were not the case, if your background music wandered from key to key or had no particular rhythm, it would stick out dramatically and would thereby defeat the very purpose the music was meant to fulfill— to *supplement* the message, not to divert the listener's attention. (Keep in mind that we're dealing here with music that's being used in a supporting capacity, and not as a solo entity.) Similarly, if your editing disturbs this rhythmic flow, your audience will notice it before you can say "oops."

Do you tap your foot or drum your fingers when there's music playing nearby? If so, you've probably noticed that your tappings, like the beats in the music, are very regular and predictable, so much so that whereas you may think you're tapping along with what you hear, in fact you're tapping along with what you *anticipate*. Before a beat sounds, your foot or finger is already in motion, rising, and then falling, so that when the beat does sound, your own percussion will sound simultaneously. The fact that our rhythms are so regular makes editing music easy.

Our music contains strong and weak beats, which function like stressed and unstressed syllables in our language. If you're not sure where the strong beats are, watch your foot. Chances are that you're tapping on the strong beats (downbeats). Between them are weak beats (upbeats). Read (or sing) this line aloud:

Camp-town la-dies sing this song

If you were tapping out this rhythm, your foot would fall on the downbeats (Camp-, la-, sing, song) and rise on the upbeats (-town, -dies, this). When you tap along with the strong beats, it's called tapping *on the beat*. If you're tapping on the weak beats, you're tapping *off the beat*.

Duke Ellington once remarked that white folks seem to tap on the beat, and black folks tap off the beat. I don't know whether he was being facetious, but this theory does tend to account for the fact that black music (ragtime, jazz, rhythm & blues) generally has a lot more *swing* to it than white pop or folk music (By *swing* I mean a catchy, irregular rhythmic feel, resulting from the use of any number of devices: syncopation, musical triplets laid over a two- or four-beat accompaniment, etc.)

Try this. Sing "Camptown Ladies" again, but this time, tap your foot on the downbeats, and clap your hands on the upbeats, as shown in the accompanying figure.

Camp -	town	la -	dies	sing	this	song
tap	clap	**tap**	clap	**tap**	clap	**tap**

See how the clapping on the weak beats does tend to make the tune swing a little more?

If a *beat* is the basic unit of musical rhythm, the next step up is the *measure*, a grouping of a few beats (almost always 2, 3, 4, or 6), which repeats throughout the composition. Unquestionably, the most popular rhythm pattern is 4/4; four beats per measure, with each quarter note receiving a whole beat. This time signature was (and is) so popular, that it is referred to as *common time*, and is indicated on most sheet music by a capital *C* adjacent to the key signature (sharps or flats) on the first staff. However, though four-beat time is dominant, there are others that you can easily recognize. A polka has a two-beat pattern. Everybody's least favorite beginning piano piece, "Chopsticks," has a six-beat pattern. If you play a waltz, you'll hear a distinct, repeating pattern of three beats (1-2-3, 1-2-3, 1-2-3, etc.), with the first beat being strong, and the other two weak. Take the line from "We Three Kings" in the accompanying figure.

WE	three	/	KINGS	of	/	OR-i-ent	/	ARE
1-2	3	/	1-2	3	/	1-2-3	/	1-2-3

Note how the line easily breaks into measures of three beats each. Also, note that the initial beat of each measure (italicized) is strong, and beats 2 and 3 are weak (*one*-two-three, *one*-two-three).

Now that you understand beats and measures, here's Rule 1 of music editing:

If at all possible, make your edit marks on *identical* beats.

Think about it. If you make your first cut on, say, the second beat of a measure, then, in order for the rhythm to continue smoothly, the second cut should be made at the second beat of another measure. The accompanying figure shows what it might look like, editing a chunk out of a piece of music that has four beats per measure.

M5	M4	M3	M2	M1
4—3—2—1	4—3—2—1	4—3—2—1	4—3—2—1	4—3—2—1

Note that if I butt the third beat of Measure 2 against the third beat of Measure 4, the rhythm pattern (4—3—2—1) remains intact, although I've removed two full measures (8 beats).

Although you can edit to any beat, I prefer cutting on downbeats. If there's any problem with the transition, the added punch of the beat can serve to further mask the edit. This would be handy, for example, if the beats in the piece you're editing aren't particularly pronounced.

One nice thing about disco music was the fact that the beats punched out of the background so distinctly. Cuing to any beat was a cinch. However, as you can guess, not all music has this characteristic. For example, it may be tough cuing to a beat in a mellow track that has lots of violins. The beats may seem to slide one into another—nice to listen to, but a pain to edit. In cases like this, listen for the instruments that are beating the time, instruments like percussion or bass. After you make your marks, here's how to check to see that they're correct:

1. Cue the tape to your first mark.
2. Keeping strict rhythm, count aloud a measure of beats and any beats between your mark and the beginning of the measure, and roll the tape. In the preceding example, for instance, because the mark is on the third beat, I'd count "1-2-3-4 1-2-[hit tape]." If the rhythm sounds smooth for the remainder of the measure, your mark is correct. If the rhythm sounds less than perfect, then recue to the mark, and then make another mark perhaps ¼-inch to the left or right of your original, depending on whether your original mark is behind or ahead of the beat. Keep repeating this process until you have the beat marked exactly.
3. Repeat the preceding procedure with your second mark.

Believe it or not, it takes longer to explain this checking procedure than to do it. Remember, once you get the hang of editing, this procedure is really necessary only when the beats of your music are hard to cue to.

Unfortunately, when I said that you are to edit to identical beats whenever possible, the phrase "whenever possible" correctly implies that editing to identical beats is not always possible. In the previous example, we spliced the third beat of Measure 2 to the third beat of Measure 4. This was easy because both beats were *in the clear*, that is, when these beats sounded, we heard only instrumental music. However, what would have happened if a vocal line had begun on the second beat of Measure 4? Editing to the third beat of that measure would have cut into the singing. The accompanying figure shows what I mean.

M5	M4	M3	M2	M1
4—3—2—1	4—3—2—1	4—3—2—1	4—3—2—1	4—3—2—1

————————————————vocal   instrumental

Let's say that you again want to splice the third beat of Measure 2 to the third beat of Measure 4, but in this case, a vocal begins on the second beat of Measure 4. If you went ahead with your original plan, you'd lop off the first beat of that vocal. The same would hold true if an instrument such as a trumpet began playing prominently on that second beat. The best course of action would be to make your first edit mark at the second beat of Measure 2, and the second mark at the start of the vocal (second beat, Measure 4). The vocal would be intact, and the rhythm would be fine.

However, what would you do if there were a problem with the second beat in Measure 2? What if there were an audible click there, or an old splice that you recorded over? (You shouldn't ever record over an existing splice because the sound at that point will momentarily drop off. Use tape that's free of splices—or at least fast forward your tape past the splice(s) and record where the tape is intact. If, for some strange reason you have to record over a splice, make sure the tape's running at the fastest speed possible. You'll stand a better chance of minimizing the dropoff.) You could, of course, make your first mark at the second beat of either Measure 1 or Measure 3, although if you took the former course, you'd still have the click in Measure 2 to contend with. However, assuming you had no beat 2 available for you to use, what would you do? Here's Rule 2 of music editing:

> If identical beats aren't available, *similar* beats can be spliced together—that is, strong beat to strong beat, weak beat to weak beat.

In other words, if Beat 2 in Measure 2 isn't usable (or if I want to get rid of it because it's no good), and I don't have any available identical beats, I'll look for the nearest available weak beat (because I want to splice to Measure 4's second beat—a weak beat). The weak beats closest to Measure 2, Beat 2 are Measure 2, Beat 4, and Measure 1, Beat 4. (Remember, in four-beat measures, Beats 1 and 3 are strong, 2 and 4 are weak.) I can make my first edit mark at either of these two spots (although I'll certainly choose Beat 4, Measure 1 so as to cut out the click at Measure 2, Beat 2) and make my edit.

Now before the purists start yelling—yes, this type of edit has a flaw. The beat rhythm (strong-weak–strong-weak) will continue unimpaired. However, unless we remove a *complete* measure or a group of *complete* measures with our edit, we're doomed to have a part of a measure left behind. In this case, by joining a Beat 3 to a Beat 1, we'll be off by half a measure (two beats). The accompanying figure shows what it will look like.

M5		M4	M1	
4—3—2—1	4—3—[24]	—3—2—1		actual beat count
2—1—4—3—2—1—	4	—3—2—1		what you hear

As you can see, the melding of Measure 1, Beat 4 with Measure 4, Beat 2 results in a loss of two half measures. Also, because of the half measure we cut out, we have a half measure remaining. Therefore, the count we perceive and the actual beat count will be different by a half measure (two beats).

Fortunately, if you have a voice track rolling as this edit hits the ear, the chance of anyone catching on to the error is about the same as the chance of the Heimlich maneuver becoming an Olympic event. In fact, a listener would have to make a concerted effort to ignore the voice, and assiduously count the beats in order to discover what you've done. What's more, even if the edit plays in the clear, it's doubtful your audience will notice it. In fact there are a handful of classic rock-and-roll recordings that were either assembled or trimmed using similar beats as edit points, and without any adverse reaction from the listening public—then or now. Figure 10.3 shows how to choose the beats at which to make your cuts for editing.

If you need to edit a piece of music, and for some reason, no available combination of identical or similar beats can be found, there's not a great deal you can do. You'll have to resort to trickery of some kind and be content with merely covering your tracks. My favorite dodge is to put the splice anywhere, and just before the edit plays, mask it with a loud, lengthy punctuator (cymbal crash, speeding train, that sort of thing). You could also bring up an ambient sound track before the splice comes by, and fade your music. Later, after the splice has passed, you can fade the SFX and bring the music back up.

Don't get the idea that editing is solely the removal of a chunk of tape. Editing can also involve the fusion of two or more completely different recordings. Although the examples I've used to illustrate rhythm matching have involved cutting a portion from a single piece of music, the rules hold true if you have to join different pieces, even if their tempos (*tempi* for purists) and rhythm patterns are different.

When joining two pieces of music, splicing at *identical* beats is critical, and if at all possible, *on the first beat of a measure.* In most of our music, regardless of rhythm pattern, the strongest beats are usually the first beats of measures. Splicing at *similar* beats may or may not work, depending on the music tracks you're using. You also could mask the juncture with SFX, but more often than not, when I fuse differing pieces of music, I do so for dramatic effect. When a new piece of music punches in, I want the audience to hear it; masking would obviously defeat the purpose.

Key Matching

The second aspect of music editing that must be handled is that of key matching. There's not nearly as much ground to cover here because, unlike rhythm alteration, which is easy to influence with thoughtful use of a blade and splicing tape, there's not much you can do to alter the musical key of a composition. In fact, there are only two ways to alter the pitch of a piece of music:

1. play the piece back at a speed that's either faster or slower than the speed at which it was recorded (see Chapter 11, "Tricks of the Trade," the section on vari-speed effects)
2. use a digital pitch-changer (see Chapter 13, on "Processing")

BEATS PER MEASURE	BEAT #	CAN BE SPLICED TO	FIRST CHOICE (IDENTICAL)	OR	BEST SECOND CHOICE (SIMILAR)
2	1		1		2
	2		2		1
3	1		1		No second choice
	2		2		3
	3		3		2
4	1		1		3
	2		2		4
	3		3		1
	4		4		2
6	1		1		4
	2		2		5
	3		3		6
	4		4		1
	5		5		2
	6		6		3

10.6 Depending on which beat you use to make your first edit mark, if you don't trust your ears or instincts, use this chart to help you place your second mark.

My dictionary (*Webster's Seventh New Collegiate Dictionary*, G. & C. Merriam and Co., 1965) defines a musical key as a system of seven tones based on their relationship to a tonic. Some help! If you're not familiar with music, think back to elementary school. Remember do-re-mi-fa-sol-la-ti-do? These are the syllables used in a basic musical scale. If you remember the melody for this standard scale, you might also recall that you can begin the scale on any tone you choose. The tone you begin the scale with (the do note) is called the scale's *tonic note* or simply the *tonic*. The interval from one tonic to the next tonic (from low do to high do) is an *octave* (from the Latin *octo* meaning *eight*—for the eight tones). To differentiate one scale from another, each scale is identified by the name of the tonic. In other words, the key of C is a do-re-mi-fa-sol-la-ti-do scale with the note C as the first (do) note; the key of G# has G# as do; and so on.

A typical popular song (if such a thing exists), based on Western conventions of musical theory, will tend to remain in one key for its duration. An alternate is that it may remain in one key for most of the piece, and change smoothly into another key (modulate) as a dramatic touch near the conclusion. If you listen to Frank Sinatra singing "New York, New York," you'll note that he sings most of the song in one key but modulates near the end, just before "these little town blues."

We don't like our music to jump erratically from key to key. That's why

Table 10.1   Key Chart

KEY	FIRST	SECOND	THIRD	FOURTH
A	A	D	E	F#/Gb minor
A#/Bb	A#/Bb	D#/Eb	F	G minor
B	B	E	F#/Gb	G#/Ab minor
C	C	F	G	A minor
C#/Db	C#/Db	F#/Gb	G#/Ab	A#/Bb minor
D	D	G	A	B minor
D#/Eb	D#/Eb	G#/Ab	A#/Bb	C minor
E	E	A	B	C#/Db minor
F	F	A#/Bb	C	D minor
F#/Gb	F#/Gb	B	C#/Db	D#/Eb minor
G	G	C	D	E minor
G#/Ab	G#/Ab	C#/Db	D#/Eb	F minor

many key changes we hear today involve a small shift upward or downward of only a half-tone (from C to C# or F# to G, etc.). If you splice two pieces of music together, each being in a different key, even though you pay close attention to splicing at identical beats, the odds are that the jarring shift from key to key will be only slightly lessened by the smoothness of the rhythm transition.

If you're going to splice together two pieces of music, it would be best to join compositions that are in the same key. In fact, it won't sound right unless the two pieces are in identical or compatible keys. However, some combinations of keys, while not identical, are compatible and sound pretty good, although not as good as identical keys. Table 10.1 shows a chart listing all the keys, and combinations of keys in the order of preference.

Thus, if you have a piece of music in the key of D, your first choice would be to splice it with another piece in D, the second choice would be a selection in G, and the third choice would be a selection in the key of A. In addition, a piece in B minor would also work. (Although most of us tend to think of music as having two basic modes, major and minor, each scale actually has seven modes, each with its own characteristic sound and effect on a listener. Check a good music reference [like *Grove's Dictionary of Music and Musicians*] for more information on musical modes.)

The question most likely to be on your mind right now is "How do I determine the keys of the music in my library?" There are a number of ways:

1. If you're lucky enough to know someone with perfect pitch, grab him/ her. People with perfect pitch can instantly tell you what key a piece is in.

2. If you're musical, bring a small keyboard into your studio. Even one of

those kiddie keyboards with minikeys will do, as long as the pitch is true. Play a piece of music, hum the tonic (the do note), and find that tone on your keyboard. That's the music's key. Repeat this procedure with the rest of your library.

3. If you're not musical, find someone who is, and pay him/her to do the preceding job for you. Seriously!

At the radio station where I was production manager, it took my colleague, Peter King (who has perfect pitch) and me nearly 6 weeks to determine the keys to all 1100 songs in our library, as well as to all our jingles and our production music. On each music cart, we affixed a label indicating the key the music began in and the key in which it ended, in case there was a key modulation somewhere in the middle. A typical label might look like the accompanying figure.

$$A^b \ / \ A$$

This would tell me that the song on this cart began in the key of A^b, and ended in the key of A. A small label like this went onto all our music and jingle carts.

Once the *keying* and labeling was accomplished, it became necessary only to make sure that any new music coming into the station was labeled by key before reaching the control room or the production studio.

Keying the music accomplished a number of things. First, it made the use of multiple music tracks in a single spot less likely to be irritating to the listener. Construction of medleys was much easier and infinitely more pleasing to the ear. (Do you realize how many songs John Denver recorded in the key of G?) On the air, our announcers had a field day with the system. Whereas competing jocks put their record sweeps (groups of records played consecutively with no breaks between them) together with hunches and hope (or merely at random), our guys used the key chart to ensure that their music sweeps sounded *great*. Our audience had no idea why our music mix sounded superior. After all, other stations in the market were playing the same songs. We were, in effect, playing a mind game with our listeners, packaging our music in such a fashion so as to capitalize on the desire of our ears to hear adjacent pieces of music in identical or compatible keys.

Just as you had to camouflage a splice between dissimilar beats, joining chunks of music whose keys are completely incompatible is a crapshoot, and the juncture should always be masked by either a voice track or SFX.

If keying your music and using this system seems time consuming and a bit awkward, I won't argue the point. However, remember that success in radio and in production is often a case of pleasing your audience by outthinking and outdoing your competition. Even if your production and announcing people use the key chart only half the time, you'll still be a leg up on the folks across town. In a game of inches, this innovation can represent feet.

Multitrack Editing

Before we leave the subject of editing, you need to know that your multitrack recorder can save you much cut-and-splice time with a technique known variously as *jumping tracks* or *skipping tracks*. Here's how it works.

You're reading and recording your voice on a track, let's say Track 1. You trip over a word. You could simply let the tape continue rolling, reread the line you flubbed, and, when you're done, rewind the tape and edit out the mistake. Instead, you could do this:

1. After you've goofed, switch Track 1 into the sync or the sel-rep mode.
2. Rewind the tape a few feet before the mistake.
3. Change the record select from Track 1 to Track 2.
4. Roll the tape.
5. Monitor Track 1 through your phones. When the sentence or phrase in which you tripped comes along, start reading from that point and recording yourself from that point onto *Track 2. Note:* Make sure Track 1 is NOT recording onto Track 2 (cross-printing).
6. Finish your read (on Track 2), and rewind the tape to the beginning.
7. Play Track 1. At the point where you began reading on Track 2, and using only the key switches, *instantly* close off the pot for Track 1 *and* open the pot for Track 2. As long as the two tracks are at the same loudness level, the switch from track to track will be imperceptible. In effect, you've eliminated your mistake without an edit!

Let's purposely trip over "Mary had a Little Lamb," as shown in the accompanying figure.

	X	
(cough) saw eceelf sti ,bmal elttil a dah yraM		Tk. 1
.... wons sa etihw saw eceelf sti		Tk. 2

On Track 1, you coughed in the middle of a line. So you wind the tape back, listen to Track 1, playing back through your headphones (in the sync or the sel-rep mode), and start recording live onto Track 2, beginning with the sentence you coughed in. When jumping tracks, it's a good idea to pick up the read from the beginning of a sentence or phrase or line, rather than in the middle of a sentence, phrase, or line, because at the beginning, there's invariably a tiny break during which you can switch tracks.

Rewind the tape, and listen to both tracks, adjusting the playback levels of both pots to ensure that they're identical. Rewind again, and play the tape, with the Track 1 pot open, and the Track 2 pot closed. We hear the voice on Track 1 say "Mary had a little lamb." At the pause after the word *lamb* (X marks the spot), simultaneously shut off the key to Track 1, and open the key to Track 2. You'll hear the voice continue, "Its fleece was white as snow . . . " without a hitch.

If you're uncomfortable flipping both key switches together, you can open the pot to Track 2 a second or two before lamb. That's assuming there's *absolutely* no noise on the track prior to the voice's pickup. If the track's clean, you can leave it open, and just snap the pot to Track 1 closed after "lamb."

The only thing to guard against, aside from different volume levels on the different tracks, is the chopping off of sound when, in the preceding example, Track 1 is shut off. You don't want to clip the end of lamb.

Let's back up a bit. Assume you've made your error on Track 1, and you've picked up the read on Track 2. If, after a while, you stumble again (this time on Track 2), rewind the tape a bit, and repeat the process, except this time, pick up your read on Track 1. Track jumping is invaluable if you have a lengthy (5 minutes or more) piece to read. You can jump back and forth between two tracks and never touch your editing block.

I know that producers are an odd lot. Honestly, there are few things in this business that can bring a smile to a tired face as readily as a perfect edit. Perhaps it's because the perfect edit is undetectable, so it's like being on the inside of an inside joke. At any rate, the ability to cut and patch bits of tape into a smoothly flowing final work is what separates the pros from the pack.

Practice!

Review

1. Editing is used to remove errors, join separately recorded pieces of tape, and create special effects.
2. Edits must be perfect and undetectable.
3. Editing tools consist of a splicing block, a single-edge razor blade, a grease pencil, and splicing tape.
4. Make your grease marks on the tape at the playback head (or record head if you're in the sync mode). Avoid getting grease on the head itself.

5. Use the diagonal (45-degree) groove on the splicing block for most of your edits. The joint will be stronger.

6. A good edit should maintain the natural flow of the language.

7. Avoid touching the oxide side of the tape unless you're handling a segment of tape that's going to be removed.

8. Every now and then, demagnetize your blade.

9. The vertical groove on the block can be used for tight editing between words, or as a reference for editing without a grease pencil.

10. When a pair of edit marks are some distance apart, use the *edit* feature of the tape deck to spool the tape off from the supply reel.

11. Try to edit at explosive or hard consonant sounds, even if they're in the middle of a word.

12. When editing music, a splice must never cause a break in the rhythm. Edit at identical beats, or if that's impossible, edit at similar beats. If all else fails, mask the juncture with a punctuator of some kind.

13. When splicing two pieces of music together, make sure the pieces are in the same key or at least compatible keys.

14. *Jumping tracks* allows you to edit without cutting the tape by combining good portions of two tracks onto a third track.

11

Tricks of the Trade

Advanced Techniques, or Who Says Special Effects Are Only for Movies?

Audio producers are a lot like magicians in that they tend to guard their secrets jealously. After all, if you're in a competitive market, why divulge something that could give you an edge over your rivals?

I must confess that although I stumbled onto many of the tricks in this chapter without outside help, I'm sure that there are other producers who've stumbled across the same tricks. We are a curious lot, constantly asking ourselves, "What would happen if I did this?" or "What would I get if I mixed this and that?" Most discoveries in the studio are accidental, and, as such, it's only logical to assume that someone else has had the same accident. I've never seen anyone else use many of the methods in this chapter, but I can't in good conscience claim to have invented any of them.

One word of warning: experimentation is the cornerstone of creative production, but keep in mind that you're dealing with sensitive, *expensive* electronic equipment. Be gentle, and be aware that there's a difference between pushing your gear to the limit and abusing it. I've repeatedly stressed the care and maintenance of your tools, and I want to reemphasize this point as we get set to head into new and often strange territory. Occasionally, I may put a grease mark on a tape recorder or turntable or console as a reference of some sort, but I always remove it upon completion of the job. If you don't respect your tools, you can't possible respect what they can do for you, and if that's the case, I'd recommend looking for another line of work.

Basic Tape Recorder Effects

You can always tell creative producers from noncreative ones. Creative producers find atypical uses for equipment. For example, take the tape recorder, the single most important piece of equipment in the studio. The manufacturer intended the following—that you would

1. place reels on the deck
2. thread the tape in the specified way
3. push/turn the appropriate buttons, knobs, or switches to make the machine record or play back, rewind, or run fast forward

Can any of these actions be varied? Can any of them be bent, twisted, or modified in some fashion, so as to produce something the manufacturer hadn't intended? Absolutely! See the section of this chapter on tape loop, for instances, where no reels are used. Threading the tape in a rather unorthodox manner, will. cause it to run backward, as described later in this chapter. As for the standard functions of the deck, you can do a number of things that you won't find in the operator's manual.

The Invisible Edit

What do you do when a spot comes into your station on a reel, and, when carting it up, you notice that the music at the head wows in? What if you're using a disc, and you want a small excerpt from the middle of a band, and it wows in even if you slip-cue? You have a number of options: (1) standard edit, (2) running start, and (3) invisible edit.

Standard Edit

With the defective spot, a simple edit should suffice. However, be careful that once the offending wow has been removed, the music doesn't jump in awkwardly. If the music does sound funny, you might try editing tight to the voice. Even if the music isn't perfect, the voice may lessen the problem. Back the disc up a half-revolution, slip-cue, record, and edit out the unwanted music. The extra half-turn not only gives you something to cut but may also provide you with a pickup to your music if you're lucky.

Running Start

This will work with either your reel or a disc. It's not difficult, but it does require split-second timing. Let's deal with the reel first. Cue up your tape just past the wowing, and make a grease mark on the top of the reel, near the edge (Figure 11.1). Try to make your mark opposite something on the deck that you can use as a reference (i.e., a screw head, another grease mark, etc.). What you're going to do is close the channel off with the key switch, back the tape up, roll the tape, and open the key when your grease mark reaches the reference point. To simplify things, make your mark at the top of the reel (twelve o'clock). Back the tape off two full turns (or more if you like). Close the key. Roll the tape. Watch for your mark, and the instant it completes its

11.1 Grease mark just past the wowing.

second (or whatever number you've chosen) revolution, open the switch. Another option is to mark the tape itself and then open the key when the mark hits the playback head. If the music sounds funny jumping in like that, follow the same procedures as with the standard edit—that is, make the spot start right at the voice.

With a disc, you can also use the running start. Cue the record to the point you want, and *note the position of the label!* Reference the label's position against something else on the turntable, or make a small grease mark on the turntable outside the edge of the disc (don't ever make grease marks on a disc), or note the clock position of the top of the label or some other feature. Back the disc up two turns, close the key switch, roll the turntable, watch the label, and open the key at the appropriate moment. If you choose to put a grease mark on the gear, make sure you clean it off when you're through.

Invisible Edit

This is a great way to solve your problem. It requires you to make use of the tape recorder in a way that few production people realize is possible. What you're going to do is to use the erase head as a tool in its own right. This technique can be used for repairing a tape that wows at the head and for cleaning up an excerpt from a disc.

11.2 The mark is rolled back to the left of the erase head.

Repairing a Tape

1. Make a grease mark just past the wowing (see Figure 11.1).
2. Roll the tape back until the mark is well to the left of the erase head (see Figure 11.2).
3. Lift the tape out from between the capstan and the pinch roller (see Figures 11.3 and 11.4).
4. Close the pot or put it into cue mode to avoid feedback, put the recorder into the record mode, and roll the tape (hit the play/record buttons) while your left hand exerts enough pressure on the left reel to keep the tape from moving.
5. Keeping your eye on the erase head, manually move the tape ahead until your grease mark is at the center of the head (see Figure 11.5). Then back the tape up a foot or two (see Figure 11.6). What you've done is use the erase head to eliminate everything before your mark, including the wowing music.
6. Stop the tape recorder, replace the tape between the capstan and the pinch roller, put the deck back into the play mode, open the pot, and your tape is ready for use.

11.3 The tape is lifted from between the capstan and the pinch roller.

11.4 The tape is clear of the capstan and the pinch roller.

11.5 The tape is rolled while the left hand exerts enough pressure on the left reel to keep the tape from moving. Move the tape ahead until the mark is at the center of the erase head.

Cleaning Up an Excerpt from a Disc

1. Cue the disc to the point where you want the excerpt to start.
2. Back the disc up a half turn.
3. Record the disc from this point onto a reel. Then follow the forgoing procedure for repairing the head of a tape.

What you're doing is using the head to do your editing for you. The whole process won't take you more than 15 seconds, from the time you make your grease mark to the time you play back your tape. That's considerably less time than if you had opted for either of the other two methods. The invisible edit is quick, clean, accurate, and uncomplicated—in short, a marvelous production

11.6 Back the tape off, erasing everything before the mark.

tool. Most people regard the erase head as strictly a means to an end, a device that readies the tape for recording. Here, we actually make direct use of it. I remember when chicken wings were looked on as useless (except to the chicken). People often threw them away. However, some enterprising individual decided to cook them and serve them with hot sauce, and voila! Profit from waste. That's just what you've done with the erase head here. *Nothing is useless. Everything can be made to serve your purpose, given the right set of circumstances!* No matter how innocuous a sound or device appears to be, you can always find a place for it.

For example, how about the garbled sound you get when you run a tape with speech on it across the playback head in rewind or fast forward mode? Useless? Not at all. How about this commercial application:

1   V1: Waldo's Widgets are now on sale.
2   Get a Left-handed Extrapolation Widget for only $19.95.
3   V2: Did you say $19.95?
4   V1: I sure did.
5   V2: Are you positive you said $19.95?
6   V1: Let's check.

[:01 of garble dubbed in]

7   [Redub Line 2: "Get a Left-handed Extrapolation Widget for only $19.95."]

[:01 of garble dubbed in]
8   V2: I guess you did say $19.95. . . .

The effect is that the characters are able to roll themselves backward and forward in time, and you've been able to create the effect by using an ostensibly useless sound. Speaking of using useless sounds, listen to John Lennon's opening riff on "I Feel Fine": feedback!

Miscellaneous Tape Recorder Effects

Here are a few other unorthodox ways you can use your tape recorder:

Pitch Change

If your machine doesn't have a variable speed control, you can make the pitch dip and rise slightly by applying pressure with your thumb on the left reel (see Figure 11.7). Record "I feel sick," and stretch out the vowel when you say "sick." Play it back, and press on the left reel when you come to "sick": The pitch dips. When you release the reel, the pitch rises again, giving an odd nauseous sound to the word.

Tape Loops

Loops can be used with either ambient sound effects or music. The idea is to take a few seconds of something and make it last considerably longer, for example making a few seconds of background noise, or a short musical phrase last for :30 or :60 seconds. One way to do this would be to record the piece you want to stretch a few times, and then edit all the recordings together. Unfortunately, many edits do not always make a smooth tape. If your sound or music is at least :06 seconds long, you can make a loop. With both music and ambient sounds, the longer the duration of your initial sound, the better, because with a short loop the sound may repeat so many times as to become noticeable. Let's first deal with an ambient sound.

Ambient Sound Loops

1. Dub your sound once (or twice if necessary) onto a reel.
2. If you've had to dub twice, edit the two together.
3. Make a cut at the head of the sound and at the tail, and splice the two ends together.
4. Remove the reels from the deck, and thread the loop, remembering which end is up; otherwise it'll play backward. As long as the loop is threaded between the capstan and the pinch roller and around the take-

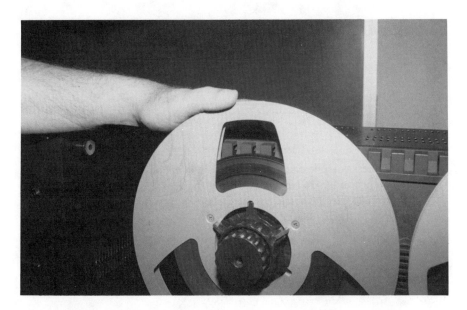

11.7 The poor person's pitch change.

up gate/guide that turns the motor on, when you play the loop, it'll go around and around forever (or until the world's energy supply runs out).

The only problems with loops are with the edit and/or the length. If the head of your sound is louder or softer than the tail, the sound level will jump at the splice. If at all possible, make your cuts where the loudness levels are identical or nearly so. In regard to the length of the loop, too long is better than too short. As long as the loop can encircle the head stack and the take-up gate, you'll be fine (see Figure 11.8). If the loop won't reach the gate, manually hold the gate up to engage the motor, or put something against the gate to prop it up (a tape dispenser, a book, a small building, etc.). The important thing is to maintain enough tension so that the loop runs smoothly across the playback head (see Figure 11.9).

If your loop is very long, don't be afraid to stretch it over a couple of other machines, or across the room, or out the door if necessary (see Figures 11.10, 11.11, and 11.12). I once had to use a 12-foot loop. I had an assistant standing outside the studio holding a pencil with an empty reel spinning on it. The loop went from the tape deck, out the door, around his reel, and back. It looked silly, but in production, I'm more concerned with how something sounds than with how it looks.

To maintain proper tension you can also let the loop hang down in front of the deck, and place an empty reel in the bottom of the loop for stability (see Figure 11.13). If you do this, make sure the portion of the loop that's hanging

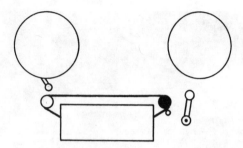

11.8 The smallest possible tape loop has to at least encircle the headstack and the capstan and pinch roller. To make it move, put the deck into the edit mode, and hit the play button. In the play mode, for the capstan to turn, you'll have to somehow prop up the take-up gate (idler).

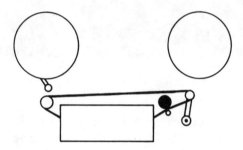

11.9 A slightly longer loop.

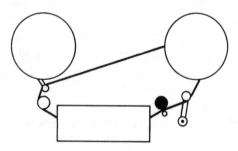

11.10 A loop may pass around one of the reels.

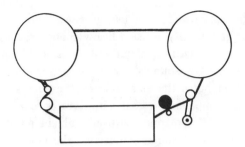

11.11 A loop may also pass around both of the reels.

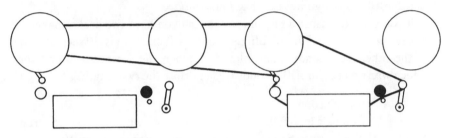

11.12 A loop may also pass around more than one tape deck.

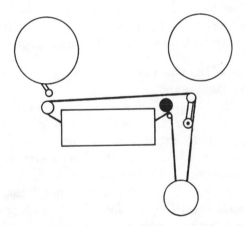

11.13 Tension in a long loop can be maintained by placing a reel in the loop.

hangs between the capstan and the take-up gate; it will not only keep the tension but also keep the gate up if you use a heavy enough reel as an anchor (such as a 10-inch one). One final note about ambient sound loops: listen carefully for any speech or distinctive sound(s). If everything is random, the listener won't notice any repetition, but if there's a voice on the track saying something, or some other obvious sound, and it comes up three or four times, it'll stick out like bad plaid. I heard a spot once with an ambient restaurant track in the background. However, it became obvious that the producer had used a loop for the background, because every four or five seconds I heard a distance voice call out, "Waiter . . . Waiter . . . Waiter . . . " Either this was a loop, or the track was recorded in a restaurant that had lousy service.

If you want to ensure a random sound, and you have two decks, make two loops of the same sound but of different lengths. If one loop is :10 seconds long, and the other :11 seconds, when played together, the composite sound they produce won't repeat for :110 seconds (i.e., multiply the two lengths).

Music Loops

A music loop follows the same rules but with one very important difference: the edit must be done in such a way as to maintain the beat of the music. See the chapter on editing (Chapter 10) if you're unsure how to do this. In general, I try to avoid short music loops because the listener can frequently notice the repetition. The only time short music loops work well is when repetition is expected, such as with rhythm tracks or drum rolls.

Threading the Tape to Run Backward

The capstan and pinch roller are designed to pull the tape from left to right. However, if you thread the tape in an "S" or "2" pattern around the capstan and pinch roller, the machine will run backward (see Figure 11.14). Fast forward and rewind functions are also reversed. Beware, though, that capstan and pinch roller placement varies from machine to machine. Also, although both of these configurations will cause the reels to roll backward, frequently only one of the threadings will keep the tape up on the heads, producing playback. I've found that the "2" seems to work well most often.

"Yes," I hear you say, "that's a cute trick, but is there any practical use for it in production?" Absolutely. I use it in any number of echo effects (described later in this chapter), backtiming, cross-fading, etc. It can also be a bonafide part of your copy:

1 SON: What is this stuff? It's awful!
2 MOM: It's a family recipe. It's been handed down for years.
3 SON: Which is why most of the family hasn't survived for years.

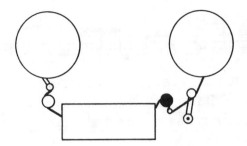

11.14 Reverse threading.

4   MOM: What a terrible thing to say. You take that right back.
5   SON: If you insist.
         [*Insert :03 of backward speech, or redub Line 3 backward,*
                  if you're a stickler for accuracy.]
6   MOM: Now don't you feel better?
7   SON: Only because I stopped eating.

Years ago, on *The Tonight Show*, a guest sang a short tune backward. Of course, it was totally unrecognizable, but when they reran the video in reverse, it turned out he had been singing "Row, Row, row your boat, gently down the stream. . . ." It was hilarious! I do the same stunt when I demonstrate reverse threading for my students. In case you want to try it, here are the words:

Meerd a tub zi file
Eelerem eelerem eelerem eelerem
Meerst a thwod ilt nedge
Tobe roy war war war

The clincher of course is to sing it! If you're at all musical, it's quite simple, once you get the awkward pronunciation down. Figure 11.15 shows how it goes.

The usefulness of this effect lies in the bizarre sound you get when you play already reversed speech backward. It's hard to describe, but if you're looking for an other-worldly type of speech, this may fill the bill. The procedure is as follows:

1. Record your line(s) naturally.
2. Run the tape backward a number of times, until you can mimic the reversed speech.
3. Record yourself speaking the line(s) backward.
4. Run this second tape backward, and listen to the sound. Whenever I want the sound of an alien speaking English, I use this effect. Sometimes I'll speed up or slow down the tape to give the voice some addi-

MEERD A   TUB ZI FILE   EE-LE-REM EE-LE-REM EE-LE-REM EE-LE-REM

MEERST A THWOD ILT NEDGE   TOBE ROY WAR WAR WAR

11.15 .

tional character or size (higher pitch— little alien; lower pitch— big sucker!).

It definitely takes some practice before your backward talk is intelligible. Here are some hints: elongate your vowels, and shorten your consonants, especially if you're singing backward. Remember, you're reversing not only your pronunciation, but also the cadence of the language. The result is a really strange effect. In addition, you can see how much simpler this effect is, using reverse tape threading, as opposed to pulling the reels off and changing spindles every time you want the track to run backward.

All of these effects require only a reel-to-reel deck—full-track, stereo, multitrack—It doesn't matter. You don't need special equipment. You do need, however, some creativity (which you already have) and some curiosity. Vary these effects, and those that follow, to suit your own needs. Experimentation is a cornerstone of creative and innovative production. Don't let your creativity be limited by your equipment: Make your equipment s-t-r-e-t-c-h to fit your ideas.

Echo Effects

You can use your equipment to create a number of echo effects including: (1) simple echo, (2) simple echo as reverb, (3) reverse echo, (4) ultrafast echo, (5) gated echo, and (6) two-deck echo.

Simple Echo

The basic echo effect is generated by recording on a machine while its monitor control has been set to play back (see Figure 11.16). Decks vary, and the playback monitor control may be labeled *tape*, *repro*, or *play*. Similarly, the record monitor control may be labeled *source*, *input*, or *rec*. If you put your deck into record mode but set the monitor to the playback mode, and at the same time, you open the pot for that deck on the console, then anything you record on the deck will (a) imprint on the tape, (b) play back from the tape through the console, (c) rerecord back onto the tape from the console, (d) play back again, (e) record again, (f) play again, etc. The sound travels in a loop from the deck to the console to the deck to the console, and so on at a rapid rate, producing a repeating echo.

If you examine the heads, you can see why this is happening. The sound imprints at the recording head, the tape travels to the playback head and plays the sound back through the console, then the sound coming through the console reimprints at the record head, and the cycle repeats.

The speed (or rate) at which the echo pulses depends on the *speed of the tape*: the faster the tape is moving, the tighter and more rapid the echo. The duration (or decay time) of the echo is determined by how far the pot is open on the console. If the pot is slightly open, the echo will fade away quickly. If the pot is well open, the echo will not only continue indefinitely, but will continue to get louder, until the tape saturates and the sound is reduced to chaos.

Try this: Don't record anything on your tape. Just set the deck to record, roll the tape (as if you were recording), and set the monitor control to play (or to repro or tape). Open the pot for that deck on the console, and slowly crank it up. Long before the pot is even half open you should begin to hear a pulsing, hissing sound. The more you open the pot, the louder the noise. What you are hearing is the faint electronic noise in the system being amplified after having been recorded and played back and recorded and played back at a rapid rate, over and over again. Incidentally, this rather disquieting sound can be used as a punctuator, or as something akin to an electronic drum roll, constantly building toward whatever you want to use as a climax—including silence. Silence, as you remember, is a great punctuator.

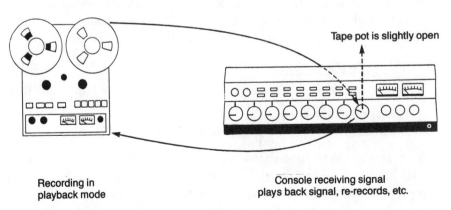

Recording in
playback mode

Console receiving signal
plays back signal, re-records, etc.

11.16 Producing a simple echo.

Simple Echo as Reverb

Although most people tend to use the terms *echo* and *reverberation* interchangeably, they are not the same. *Echo* is a distinct, discrete repetition of a sound, whereas *reverberation* is a composite, a sonic mixture of echoes bouncing around in a space. Reverb is what gives a room its presence. When you listen to a speaker, the sound you hear is a mixture of direct (from the speaker straight to you) and reflected (from the speaker off the walls, ceiling, audience, etc. to you) sound. There are electronic reverb devices on the market (see Chapter 13 on "Processing"), but if you don't have one, the sparing use of simple echo can give you a semblance of reverb.

In fact, our simple echo is very much like reverberation, in that it consists of many echoes that combine and eventually decay to nothing. If you want to simulate an empty room, slowly crank up the pot on the console while recording with simple echo, until you get the sound you want. If you want the sound of a particularly large room, back away from the mike a couple of feet, and speak louder, as if you're trying to communicate with someone at a distance, and add your echo to this. In general, let common sense guide you. The larger the room, the further away from the mike you should be. The emptier the room, the more open the pot should be.

Here's a more important use for simple echo used as reverb: when you mix voice and music tracks, you are combining tracks recorded in different locations and having little acoustic similarity. As a result, the voice track from your studio, and the music track from someplace else may not sound natural when you mix them together. The trick is to give them something in common—a touch of reverb— and I mean a *touch*. The reverb level you use here should be almost inaudible, just enough to smooth the mix and no more.

Here is one other hint for mixing any echo effects: don't use headphones unless it's absolutely necessary. The echo or reverb effects will sound much louder through your headphones than they will coming through somebody's radio. If you want to hear what your audience will hear, take off the phones.

Reverse echo

Somewhere, it's written—the Immutable Law of Echoes says that

1. First comes the sound.
2. Echoes follow, each at a progressively diminishing level of loudness.

That's the way echoes have always acted. However, a creative producer would invariably say, "I know that's the way it's always been, but what if . . . ?" Can the echo precede the sound? You bet! With sound effects, music, speech, anything. Sound effects are easiest if you use a sound that trails out nicely—a cymbal crash, for example. Just record it, play the tape backward, and you have an interesting effect (even though there's really no reverse echo). If the sound doesn't trail out naturally

1. Cue up your sound (let's say it's on a disc).
2. Roll the tape recorder in the record mode (as in the simple echo), with the tape monitor on play, and the pot open on the console.
3. Dub the sound onto the tape. You may need to try it a couple of times before the trailing echo is the way you want it. If it trails for too long, close the pot a bit. If it's not long enough, open the pot more. Remember, the echo's duration is determined by how wide the pot is open.
4. Run the tape backward.

Actually, this isn't really reverse echo. The echo is still on the tail of the sound: It's just that the whole thing is playing in reverse. To put the echo on the *head* of the sound, you'll need two decks:

1. Record your sound onto Deck 1. The head of the sound should be clean and sharp, with as short an attack time as possible (no wowing).
2. Play the sound backward. (You might as well use reverse threading. Make sure your threading keeps the tape against the heads.) Record this sound, with echo, onto Deck 2. The echo is now on the head of the sound.
3. Reverse the tape on Deck 2, and play it back. The sound is now playing in the correct direction, but the echo is reversed in its condition (starting low and building to a crescendo) and position (tacked onto the head).

If you don't want to use two decks, two tracks of a multitrack deck will do just fine. Just remember to keep the deck in the repro mode throughout the entire process. If you're in the sync or the sel-rep mode, you won't get any echo because there's no delay between when the sound plays back from one track and records onto the other; it's the delay that produces the echo.

Using the reverse-echo effect with music and/or voice is trickier but well worth the effort. To understand the process, let's deal with music from a disc. (If you don't have two reel-to-reel decks, dub the note onto a cart, and dub from the cart back onto the deck, with echo.)

1. Make sure the first note is a solid, fairly loud, clean sound (no background noise).
2. Dub this onto Deck 1. Edit out everything but that first note.
3. Dub this single note backwards, with an echo, onto Deck 2. Adjust the trail according to your taste.
4. Play the tape on Deck 2 backward for reverse-echo effect.
5. Mark the tape where the sound is loudest, probably at the end of the reversed tape.

6. Recue the disc, and cue up your tape.
7. Play the tape. As the sound builds, watch for your mark. When the mark hits the playback head, slip-cue the disc.

What you're doing here is recording the first note with reverse echo, and then tacking that reverse echo back onto the beginning of the music. As a result, that first note starts softly and builds and builds until it explodes into the music. Try this with Beethoven's *Fifth* and watch your "Intro to Music" instructor's jaw hit the floor. [When using reverse echo on a voice, the voice's pace should be fairly slow so that the whole thing doesn't become so full of sound as to become intelligible.]

Ultrafast Echo

As you know, the speed of the tape determines how fast the echo pulses. Assuming your tape deck has two speeds, 7-½ and 15 inches per second (ips), compare the two speeds with echo. Set up your deck as for the simple echo, and get on the mike. Listen on headphones as you record your voice with echo at 7-½ ips; then, while the tape is running, switch the deck into the faster speed and listen again. Hear the difference? Now, if the 15-ips echo isn't fast enough for you, try a 30-ips echo. "What!" I hear you shriek. "My deck doesn't do 30 ips." Neither does mine—so let's improvise. You'll need two decks (see Figure 11.17).

1. Record your voice on Deck 1 at 15 ips.
2. Recue your tape, and change the Deck 1 speed to 7-½ ips.
3. Set the tape speed for Deck 2 to 15 ips.
4. Record the voice from Deck 1 onto Deck 2 with echo. What you are doing is recording a 7-½-ips voice track with 15-ips echo.
5. Dub Deck 2 (slowed voice, fast echo) directly back onto Deck 1 (which is still set at 7-½ ips).
6. Raise the speed of Deck 1 to 15 ips. Your 7-½-ips voice track is now playing at 15 (which is how you originally recorded it), and the 15-ips echo is whizzing along at 30 ips! If you want the sound of a computer voice, record you voice speaking in a monotone, and run it through a 30-ips echo.

Believe it or not, using this same deck-to-deck technique, you can double the echo speed again, this time up to 60 ips (see Figure 11.18).

1. Record your voice on Deck 1 at 15 ips.
2. Drop the Deck 1 speed down to 7-½ (voice is now one speed too slow).
3. Dub this onto Deck 2, which is running at 15 ips. Don't add echo yet (see Figure 11.18.

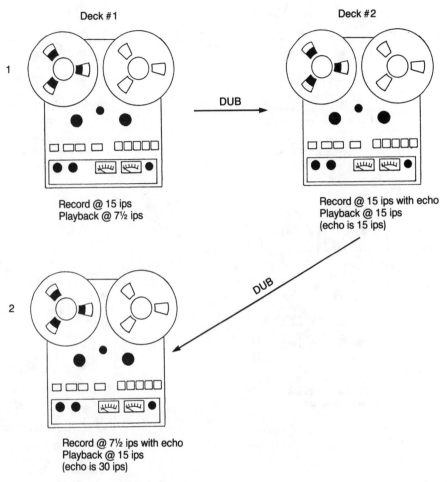

11.17 Generating a 30 ips echo.

4. Lower Deck 2's speed to 7-½ (voice now sounds two speeds too slow).
5. Raise Deck 1 back up to 15 ips (see Figure 11.18.
6. Dub the voice track from Deck 2 back onto Deck 1, with echo. Echo is now running at 15 ips, and the voice sounds like 3-¾ (two speeds down from 15).
7. Dub the voice/echo track on Deck 1 back onto Deck 2 (Deck 1 is still at 15, and Deck 2 is still at 7-½ ips) (Figure 11.18).
8. Raise the speed of Deck 2 to 15 ips. The voice is now one speed down, and the echo is pulsing at 30 ips.
9. Lower the Deck 1 speed to 7-½. Dub Deck 2 onto Deck 1 (Figure 11.18).

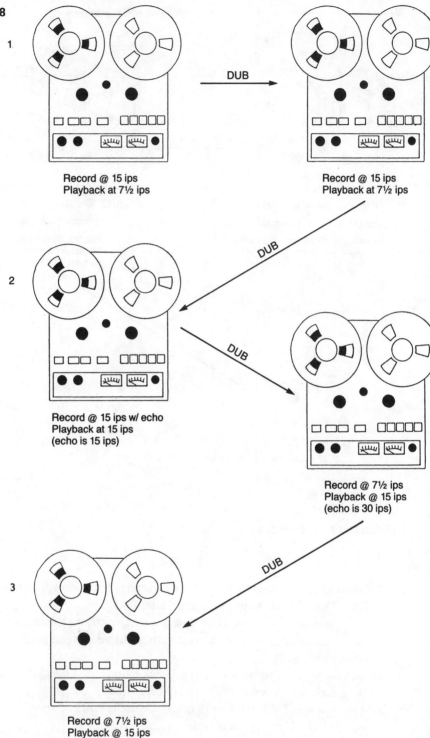

1

Record @ 15 ips
Playback at 7½ ips

DUB

Record @ 15 ips
Playback at 7½ ips

DUB

2

Record @ 15 ips w/ echo
Playback at 15 ips
(echo is 15 ips)

DUB

Record @ 7½ ips
Playback @ 15 ips
(echo is 30 ips)

DUB

3

Record @ 7½ ips
Playback @ 15 ips
(echo is 60 ips)

11.18 Generating a 60 ips echo.

10. Raise Deck 1 up to 15 ips (Figure 11.18[d]). The voice should be back to normal, and the echo is at an incredible 60 ips. Eat your heart out, Chuck Yeager!

If you have difficulty following this, here's another way of looking at it. You've taken a 15-ips voice track, dropped it down to 7-½ and then to 3-¾. That is, the voice sounds like it's at 3-¾, even though the tape on your decks never really goes that slowly. When the voice is at this very slow speed, we record it with a 15-ips echo. Then the whole thing is raised two speeds, the voice going from 3-¾ to 7-½ to 15, and the echo in turn going from 15 to 30 to 60. Got it? In addition, if you have a variable speed control on your decks and can make them run at speeds other than 7-½ and 15, you can make your echo pulse at any speed you want, not just 7-½, 15, 30, 60, and so on.

You have to really love an effect to go through a process such as this to generate it. Also, because you're losing a number of generations of tape quality (in this case, four), the number of applications is limited. However, it does give you an idea of how far-reaching your effects can be if you use your imagination.

Gated Echo

Gates are devices that, as the name implies, close off part or all of a signal. Gates are often used to cut off noises, by shutting a signal off when its loudness dips below a preset decibel level (threshold). In this case, the low decibel level of the noise becomes the threshold. Although the signal should mask the noise, when the signal is silent and the noise is in the clear, the gate instantly closes, and the noise isn't heard.

When echo or reverb is gated, the result is an interesting sound, as the echo's decay is clipped some time before it fades to nothing. Drummers frequently use gates to color the sounds from their kick and snare drums. In this instance, the gate slams shut almost immediately after the drum is hit. Even if you don't have an electronic gate, you can improvise, and you can gate your echoes using the key switch.

To do this, set up your system to produce a simple echo. I think that a fast echo sounds better gated than a slow one, so set your tape deck to 15 ips, to produce the fastest echo you can. Get on the mike, and keep your hand on the key switch to the tape recorder pot through which the echo will sound. Record yourself, with echo, saying a single syllable like "hi," and listen to the echo in your headphones. When the echo sounds the way you like it, pulsing at a nice fast clip, say "hi" again, but this time, close off the key switch a half second or so later, cutting off the echo's decay.

If you like what you hear, next, try putting gated echo on a complete sentence. You'll want to gate each word, so your hand (on the key switch) and voice must work in tandem, with the key switch opening just before each word you speak and closing almost immediately after.

For the effect to sound good, speak slowly and smoothly. There must be some space between words, space for the gated echo to be heard. So forget about trying to use gated echo on something like a Gilbert and Sullivan patter song. Not only will the echo not be heard, you'll probably break your wrist and the key switch trying to keep up with your voice.

Two-Deck Echo

Some people don't like the electronic echo we've been playing with because it doesn't sound natural enough, and you really can't control it 100%. Using a two-deck setup, you can achieve a perfectly natural echo (or at least a more natural echo), and your control is total.

1. Roll both decks in the record mode, and put a voice track down simultaneously on both machines.
2. Cue both decks to the beginning of the track, and mark on both the point where the voice starts.
3. Move the mark on Deck 2 slightly (¼-inch or so, if your deck's running at 7-½ ips) to the left. The mark on Deck 1 is still dead on the playback head.
4. Roll both decks simultaneously. Deck 2 will play a short time after Deck 1, producing a single, discrete echo sometimes referred to as a *slap-back* echo. To make it sound even more natural, lower the level of Deck 2. (The echo isn't going to be as loud as the initial sound, right?) For an odd effect, make Deck 2 (the echo) louder than Deck 1. So much for the "Immutable Law of Echoes." Some people find this effect disquieting, if not downright jarring. Rock musicians use slap-back so often that there are a number of devices on the market designed to produce slap-back on signals fed from guitars, keyboards, mikes, etc.

Creativity

Time out! Before we proceed with describing several other tricks of the radio-production trade, I want to relate a story that illustrates the point of all this information: creative thinking. This story is probably apocryphal, but it's nonetheless entertaining.

It seems that a high school physics student didn't receive credit for a test answer he felt was correct. The question went something like "How can you determine the height of a building using a barometer?" The correct answer, according to the teacher, involved using the barometer to compare the air pressure at the base of the building with the air pressure on the roof. The difference between the two was then plugged into a formula to determine the height of the building. The student, however had taken a different tack. He recommended dropping the barometer off the roof and measuring the time it took for the barometer to hit the ground. Because bodies all fall at the same fixed rate of descent (16 feet/sec/sec), it was a matter of simple multiplication to compute the height of the structure.

Anyway, the teacher marked the student's answer wrong, and that was that—except that the student protested to the head of the science department. The student and the teacher were called before the chairman, and the student asserted that the question had merely called for the height of the building to be determined using the barometer—which she had done. In fact, she insisted that there were many ways to solve the problem. For example, she said one could take the barometer up to the roof, tie a long piece of string to it, lower the barometer slowly to the ground, and then measure the string.

Another idea, she suggested, is to stand by the building and position the barometer on the ground on its end. Measure the height of the barometer and the length of its shadow. Then measure the length of the building's shadow and set up a simple proportion:

$$\frac{\text{barometer height}}{\text{barometer shadow}} = \frac{\text{building height}}{\text{building shadow}}$$

In other words, the height of the building would equal:

$$\frac{\text{barometer height} \times \text{building's shadow}}{\text{barometer shadow}}$$

Again, the problem was solved using the barometer.

Finally, the student suggested that if he couldn't get up to the roof, or didn't have any string, or didn't have a ruler to measure shadows, he'd simply go to the building superintendent and say, "If you tell me how tall this building is, I'll give you this barometer!"

I suppose, because of his logic (he had indeed found the height of the building using a barometer), the student eventually received the credit he was after. The point this story makes, however, goes to the heart of creativity. Creativity is not necessarily the invention of something brand new as much as it is (1) repackaging known material into a new form, or (2) being able to find a novel solution to an ordinary problem. The student in the story didn't invent anything new. He merely applied solutions that had solved other problems to the problem he had at hand. Instead of running through the wall, he ran around it.

If necessary, combine your effects to solve problems. I once heard a weird voice in a movie and re-created it by recording a voice with a multiple slap-back echo (see the next section), dropping the tape's speed to lower the pitch, and then adding reverse echo. If you know how to generate ten individual effects, how many new effects can you generate combining two effects at a time—or three at a time—or four—or more? The possibilities are staggering.

Production is the art of boxing yourself into a corner—
and then figuring a way out

Delay Effects

With the two-deck echo, we enter the realm of delay effects. There are a number of digital delay units you could buy, but they aren't cheap. A really good one with lots of features can cost over $5,000. Not every station manager is going to be willing to cough up that kind of money for what is perceived as a special effects toy, so if you want delay effects, you're going to have to manufacture them on your own. Fortunately, many of these tricks are easy to do. Basically, the gadgets on the market do what you did with the two-deck echo—they take a sound or voice track, duplicate it internally, and send the two sounds (the original and the duplicate) out into the console. However, instead of sending them out simultaneously, one of them is *delayed*—that is, sent out a few milliseconds after the other.

In the example of the two-deck echo, if the mark on Deck 2 is retarded only ¼ of an inch, then it would run about ³⁄₁₀₀ of a second behind Deck 1. That's 33.33 milliseconds of delay if the deck were running at 7-½ ips. It doesn't sound like much, but you'll be surprised at how different your echo will sound by varying the delay interval even a minuscule amount. I've found that if the delay is much more than a half-second (500 milliseconds), the result is a mess. The listener has a terrible time understanding what's being said, because instead of hearing sound/echo, the excessive delay sounds like sound/sound. If you've ever tried listening to two people talking at the same time, you know how difficult, as well as annoying, it can be. So beware of lengthy delays. If the delayed sound is perceived as a separate entity, and not as an echo, you've got trouble.

Flanging and fake stereo via delay are two delay effects that you may find helpful and useful.

Flanging

Flanging is probably one of the most popular of the delay effects because it's easy to do, and it produces a sound that is quite indescribable. Proceed as you did in creating the two-deck echo, but have one deck running at 15 ips and the other at 7-½ ips. When you've cued the decks up, and you've marked the head of each, start the two simultaneously without delaying one. The two tracks should combine and sound as a single track. Now, *gently* put some pressure on the left (flange) reel of one of the decks. This will cause that tape to run slightly behind the other. After a few seconds, depending on the amount of pressure you're exerting, the mixture will have a weird, hollow, swishing color to it that bears a striking similarity to Darth Vadar's voice. What you are experiencing is a phenomenon called "wave cancellation." Here's a good way to illustrate it.

1. Using two tone generators, send two identical frequencies into different pots (but through the same output) of the console, and watch the Vu meter. Set the level of each tone to –5 Vu, and make sure you set the levels one at a time.
2. When you play both tones together, if the tones are identical, the needle should rise up to and possibly above 0 Vu.
3. Now alter the frequency of one of the tones very slightly (1 Hz or less), and watch the meter. The needle will fluctuate dramatically, bobbing up above 0 Vu, then dipping down almost off the low end of the scale, then repeating the cycle.

What's happening is that instead of the two waves lining up perfectly, as they did in Step 2, now their relationship changes; sometimes the crests and troughs line up, sometimes they don't. Also, sometimes the crest of one wave (high pressure) lines up with the trough of the other (low pressure). When this happens, they, in effect, cancel each other out. By fine tuning one of the generators, you can actually freeze the needle at the bottom of the meter. Turn off one of the channels, and the needle pops back up to –5; but keep them together in this out-of-phase relationship, and the sum of the two sounds is 0!

This has to be seen (or heard) to be appreciated. Use a low frequency like 100 Hz to do this demonstration because it has a long wavelength, and is easier to freeze. The rule is this: to cancel a tone, mix it with a duplicate tone that has been delayed (or advanced) one-half a wavelength, so that the crest of one wave lines up with the trough of the other. Here's a formula to help you determine the amount of delay (*d*) necessary to cancel a given frequency (*f*):

$$d = 500/f$$

where *d* is the amount of delay in milliseconds, and *f* is the frequency of the tone in Hz. To cancel a 1000-Hz tone, mix it with another 1000-Hz tone that has been delayed 0.5 msec.

$$500/1000 = 0.5$$

Two 250-Hz tones will cancel with a 2-msec delay; 50-Hz tones will disappear when one is delayed 10 msec. A male voice that's around 100 Hz has a 5-msec cancellation interval. If this voice were recorded onto two tape decks, and the decks were played back with one 5 msec behind the other, the result would be some cancellation. In terms of actual tape length:

$$0.005 \text{ (5 msec delay)} \times 7\text{-}\tfrac{1}{2} \text{ (inches of tape per second)} = 0.0375 \text{ inches, or}$$
$$\text{slightly more than } \tfrac{1}{26} \text{ of an inch}$$

The delayed tape would have to run about ⅟₂₆ of an inch behind the other to cancel 100 Hz. Of course, the human voice doesn't dwell at a single frequency; it wanders across a whole range of frequencies. Even if the male voice in our example had 100 Hz at the center of its range, the voice would move above and below 100 Hz just as a matter of course and have lots of harmonies above 100 Hz. As a result, if our tapes have been arranged so as to cancel 100 Hz completely, frequencies on either side of 100 Hz would also be affected, although to a lesser degree. The further away from 100 Hz, either up or down, the less cancellation. So some frequencies would be totally canceled, some greatly canceled, some slightly canceled, and some not affected at all. It's this mixture that gives flanging its peculiar sound.

When we press on one of the flange reels, we are producing a variable delay that increases the longer or harder we press. Eventually, the delay would become so great as to produce a two-deck echo. As soon as you hear echo forming, release the reel, and press down on the flange reel of the other deck. The phase will slowly shift back, and, if you maintain the pressure, will slip out of phase again with this second deck providing the delayed signal. By carefully alternating pressure on the flange reels, the phase can be made to shift back and forth for a truly eerie effect. Also, flanging is best done with one of the decks running at 15 ips (or the fastest speed possible). At 7-½ ips you have to keep one of your decks running only about ⅟₂₇ behind the other to cancel 100 Hz. If the decks are running at 15 ips, this delay gap would double to a whopping (only kidding) ⅟₁₃ of an inch. I know it doesn't sound like much, but this added space gives you much more time to make the phase shift.

Whenever you attempt flanging, however, patience is the key. It may take a few tries before the frequencies begin to cancel nicely. Remember, use light pressure on the reels at all times. You're after a gradual shift, and you'll only get it with a light touch. Again, a lower frequency is easier to flange than a higher one because of the longer wavelength. As we've seen, to cancel a 100-Hz tone, one deck would have to be a little more than ⅟₂₆ of an inch behind the other. Even though that doesn't leave much room for error, compare that to the delay needed to cancel a 1,000-Hz tone:

$$d = {}^{500}\!/_{f} = {}^{500}\!/_{1,000} = 0.5 \text{ msec delay}$$
$$0.0005 \text{ seconds (0.5 msec)} \times 7\text{-}\tfrac{1}{2} \text{ (ips)} = 0.00375 \text{ inches or}$$
$$\text{a little more than } \tfrac{1}{266} \text{ of an inch}$$

The lower the frequency, the longer the wavelength. The longer the wavelength, the larger the delay necessary to cancel the frequency. Our hearing range is from approximately 16 Hz to 16 kHz, so for a 16-Hz tone,

$$d = {}^{500}\!/_{\!f} = {}^{500}\!/_{\!16} = 31.25 \text{ msec}$$
$$0.03125 \times 7\text{-}\tfrac{1}{2} = 0.234 \text{ inches or slightly more than }\tfrac{1}{4}\text{-inch delay}$$
$$\text{necessary to cancel at 7-}\tfrac{1}{2}\text{ ips}$$

For a 16-kHz tone,

$$d = {}^{500}\!/_{\!f} = {}^{500}\!/_{\!16,000} = 0.03125 \text{ msec}$$
$$0.00003125 \times 7\text{-}\tfrac{1}{2} = 0.000234 \text{ inches or about }{}^{1}\!/_{4274}\text{-inch delay}$$
$$\text{necessary to cancel at 7-}\tfrac{1}{2}\text{ ips}$$

Trying to accurately cancel any frequency, particularly a high one, is like trying to thread a very small needle, so a feathery touch is definitely the order of the day.

Fake Stereo via Delay

Stereo is the subject of the next chapter (Chapter 12), but as long as we're talking about delay effects, let's see how delay can be used to simulate stereo in a mono recording.

The main difference between mono and stereo is that with mono, the identical signal emanates from the left and right speakers; whereas with stereo, the signals from the left and right sides are different, either in part, or totally. The result is an aural image that can be positioned anywhere between the speakers to create an illusion of depth and breadth (see Chapter 12). Using two decks, you can fake a stereo image. The procedure is similar to that used in flanging.

1. Record your mono track on both decks, and rewind the tapes.
2. Mark the heads of both tapes.
3. Cue one mark up to the center of one playback head.
4. Cue the other mark to the center of the other deck's playback head, and then move the mark back ⅛-inch or so.
5. Roll the tapes simultaneously, sending the signal from one deck to the left side, and the other deck's signal to the right.

Because the two signals are not perfectly synchronized, the *wave patterns* coming from the respective speakers are not identical, and this dissimilarity is often enough to create the illusion of stereo. The difference between this method of producing fake stereo and flanging lies in the delay interval. With phlanging, the effect is generally created through a delay of 20–30 milliseconds, but with this type of fake stereo, the delay can be as long as 50 milliseconds. Why? Because according to the Haas Effect (Chapter 1), the two signals can be delayed up to 50 milliseconds before we hear an echo. Also, because we

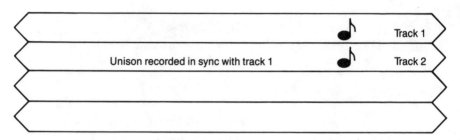

11.19 Sweetening (single unison).

 need to make the two signals as different as possible *without an echo,* we want to push right up to that 50-millisecond barrier.

If your tape's rolling at 7-½ ips, 50 milliseconds would represent ⁶⁄₁₆-inch, midway between ¼- and ½-inch on a ruler.

7-½ (ips) × 16 (sixteenths per inch) = 120 sixteenths per second of tape
120 (sixteenths) × 0.05 (50 msec) = 6/16-inch

Remember that this ⁶⁄₁₆ of an inch represents the *maximum* delay allowable before the listener hears an echo. Also keep in mind that this is *fake* stereo. Don't expect to match the separation you get with true stereo recording, although you're sure to get considerably greater depth and breadth than you would from an ordinary mono signal.

Multitrack Effects

If you have a multitrack tape deck (two tracks or more), another wealth of effects is available to you. The following six are a few I've used:

1. sweetening
2. chorusing
3. slap-back echo
4. multiple slap-back
5. stacking
6. dovetailing

Sweetening

Sometimes referred to as *voice doubling, sweetening* is the technique that Elvis (and others) used so often. What's involved (see Figure 11.19) is recording your voice (speaking or singing) onto one track, rewinding the tape, putting the deck into the sync or the sel-rep mode (see the section of Chapter 4 on multitrack recorder operation), monitoring the track through headphones,

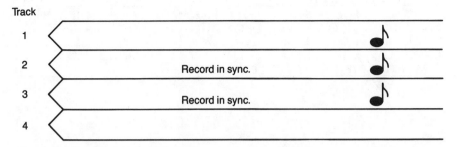

11.20 Chorusing (multiple unison).

and recording the voice again but on a separate track. No matter how careful you are, the two tracks will never be exactly the same. The pitch, cadences, attacks will always vary a bit, creating the effect of two of you speaking simultaneously.

Chorusing

Chorusing is merely sweetening carried a step further. Record your voice on at least two additional tracks (see Figure 11.20)—the more the merrier. Record them onto another deck with simple echo, and you'll sound like a one-person Mormon Tabernacle Choir.

Slap-back Echo

To get a *slap-back echo*, record your voice on one track. Rewind the tape, and dub the voice from this track onto another track, without any synchronization. Play the two tracks back, and you'll have a slap-back echo (two-deck echo) but with only one deck (see Figures 11.21 and 11.22). Do this with the deck running at a faster speed, and the echo will be tighter.

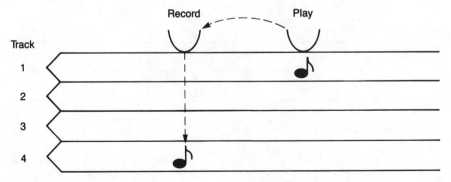

11.21 Single slap-back echo.

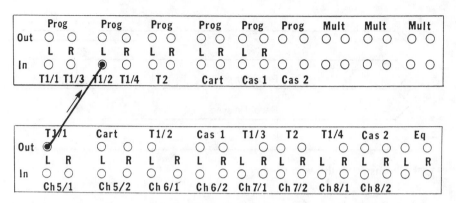

11.22 A slap-back echo can be generated without using the console by patching from one track directly into another in the play or the repro mode.

Multiple Slap-back

If your deck has four or more tracks, try this to produce a multiple slap-back echo (see Figure 11.23). Assuming you're using a 4-track machine, grab three patch cords, and patch the following (see Figure 11.24):

> Track 1 OUT to Track 2 IN
> Track 2 OUT to Track 3 IN
> Track 3 OUT to Track 4 IN

Shut off all pots on the board (except the mike pot), set the deck to record onto all four tracks, but in the playback (not the sync) mode, roll the tape, and say "Hi." Your voice will record onto Track 1, which will playback and record onto Track 2, which will playback and record onto Track 3, which will playback and record onto Track 4 almost simultaneously. As a result, if you plug your headphones into the deck, you'll hear the original "Hi" followed by three distinct slap-back echoes.

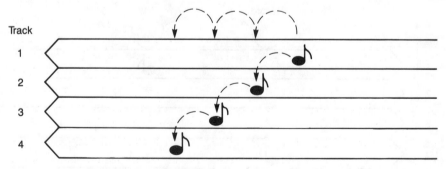

11.23 Multiple slap-back echo.

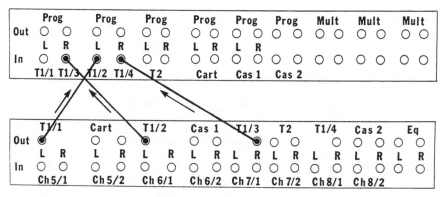

11.24 Patching for multiple slap-back.

Each track is recording and playing at the same time, recording the signal from the preceding track (which has been funneled in via a patch cord), and playing back onto the succeeding track (again, via a patch cord). The delay from track to track is of course due to the fact that when a track plays at the playback head (you're in the playback mode, remember!), the signal is recorded on the next track at the record head. Watch the deck meters, and you'll see them pop up in consecutive order when you say "Hi" onto the first track.

When you've finished, rewind your tape, open the four pots on the board with Track 1 playing loudest, and Track 4 softest, and play the tape. If you reverse the levels (with Track 4 being loudest), it'll sound something like reverse echo, only cleaner, and a little more like Max Headroom.

I wouldn't recommend multiple slap-back for a standard voice track. All those echoes bouncing around can make a voice speaking in a normal fashion unintelligible. However, for single words or sounds, this is an interesting effect.

Stacking

Sweetening involves speaking or singing in unison with a prerecorded track. *Stacking* is the same thing but involves singing harmony (see Figure 11.25). This technique was pioneered by Les Paul and Mary Ford and Patti Page back in the early 1950s, and it was continued by singers like Connie Francis, Neil Sedaka, the Carpenters, Pat Boone, and many others, especially during the formative years of rock and roll. Harmony is always beautiful but is especially so when the voices have some acoustic and/or harmonic similarity. In stacking, the voices are not only similar in resonance and harmonic structure, they're identical. With multitrack recording techniques you can make yourself into everything from a duet to an entire choir. I had a student who dubbed a Four Tops song onto one track, sang a separate line of harmony on a second track, balanced and mixed the two tracks together, and became the fifth Top.

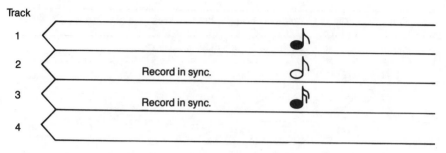

11.25 Stacking.

Dovetailing

If you can sing with yourself, you can talk to yourself. *Dovetailing* involves dialogue, with you providing both voices (see Figure 11.26). First, you're going to record all of Voice 1's lines. As you finish each line, mentally speak Voice 2's subsequent line, and continue in this fashion to the end of the copy. Roll the tape back, put the deck into the sync mode, monitor the playback through the headphones, and record Voice 2's lines onto a separate track. Rewind the tape again, play the two tracks back through separate pots on the console, and you have dovetailing. Timing is obviously critical. When recording the first track, you have to leave just enough room between lines for the other voice. If there's a little overlapping between the voices don't worry: it'll sound like natural speech. However, beware! When we recite lines *mentally*, we have a tendency to go faster than when we say them out loud, so be prepared to compensate. You'll also want to alter one of the voices so that the listener can tell them apart. Assume a character voice, or run the track through some kind of processing, like equalization or any of the echo effects you've learned— anything to make the voices distinguishable.

Bouncing

Bouncing isn't really a trick or a special effect. It's a technique that will enable you to significantly stretch the capabilities of your tape recorders. *Bouncing* (sometimes called "ping-ponging") involves combining recorded tracks onto empty tracks, either within the same deck or onto a different deck. Knowing how to bounce allowed George Martin, the Beatles' producer, to produce *Sgt. Pepper's Lonely Hearts Club Band,* with all its complexity, on nothing larger than a four-track recorder. There are two types of bounces— internal and external.

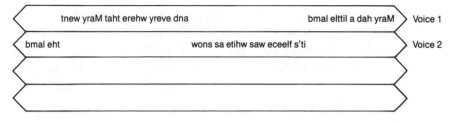

11.26 Dovetailing.

Internal Bounce

Let's say you have a four-track recorder, and you're recording a band consisting of two guitars, two singers, and a drummer. Because you should have at least three mikes on the drum kit, you're looking at seven tracks of material—three tracks for the drums, two for the guitars and two for the voices. Using internal bounces (see Figure 11.27), you'll be able to fit all these tracks onto the four-track machine you have at hand. Here's how:

1. Use Tracks 1, 2, and 3 to record the drums alone, perhaps one mike over the left side of the kit, one over the right, and one down by the kick drum.
2. Rewind the tape, and play the tracks, adjusting the levels until the three tracks are balanced to your satisfaction.
3. Record this percussion mix onto Track 4. This step represents your first internal bounce.

You've now recorded three tracks of material onto a single track. If, when you play Track 4, you're content with the result, then you're free to record new sound over the original drum tracks (1, 3, and 3). If the mix on Track 4 isn't what you'd hoped, rewind and remix. Remember, once you record over any of the old tracks, they're gone forever, and your mix will have to stand, so make sure your mix is just the way you want it to be before moving on.

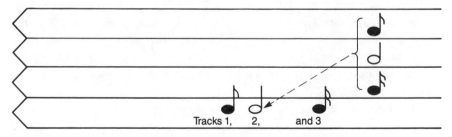

11.27 Internal bounce. Tracks are combined on the same piece of tape.

4. With the deck in the sync mode, record the two guitars on Tracks 1 and 2, in sync with the mix on Track 4. Rewind the tape, listen critically to the guitar tracks, and if they're okay, bounce them both to Track 3. You now have five tracks of material recorded onto only two instrumental tracks— three percussion tracks on Track 4, and two guitar tracks on Track 3.

5. Rewind the tape, and record the two voices, one onto Track 1 and the other onto Track 2, in sync with either or both of your instrumental tracks. Rewind and play back. You've got seven tracks worth of sound on only four tracks. This is the essence (and the incredible value) of bouncing. You'll now probably want to mix these four tracks down onto another recorder in either a mono or a stereo final mix.

There's a variation called a *modified* or *half* bounce, which involves combining recorded and live material on a separate track. In the previous example, if we had three singers instead of two, we could have done the following after recording the instrumental tracks:

1. Record one of the voices on Track 1 in sync with either or both of the instrumental tracks.
2. Put the second singer on mike live, in sync with the vocal on Track 1, and mix the two vocal lines together onto Track 2.
3. Record the third voice on Track 1.

There you have it! Eight sound tracks of material combined onto only four tape tracks.

Note: There is a glitch you may encounter if you do internal bounces on a multitrack deck. Try this on your deck, and see what happens:

1. Record a voice track (or anything) on Track 1.
2. Rewind the tape, put the deck into the sync mode, and bounce Track 1 onto Track 2.

If the bounce worked, great. However, if your deck experienced feedback, and the deck's meters pinned, here's why. In the sync mode, you're asking the record head to perform two functions—play Track 1, and record Track 2. Because these tracks are so close together on the head, they produce the same kind of feedback you'd get if you set a deck to record and also opened the deck's pot on the console or if you held a live mike in front of a speaker. It's called "Let's make 'em squeal!"

To avoid the feedback problem, you have two options:

1. Don't bounce in the sync mode from one track to an adjacent track. If, in the preceding example, we bounced from Track 1 to Track 3 or Track 4,

we'd have no feedback problem. There's enough space between the tracks on the head to avoid the problem. There's no law that says you have to bounce from track to track in numerical order. If you want to lay down music on Tracks 3 and 4, and bounce them to Track 1, go ahead. It's only with adjacent tracks that you may have a problem.

2. If you have to bounce to an adjacent track, don't use the sync mode. Use the play or the repro mode. Remember when we recorded our drummer on Tracks 1, 2 and 3 and bounced the three tracks onto Track 4? If we had done the bounce in the sync mode, the adjacency of Tracks 3 and 4 could've caused feedback. However, if we had done the bounce in the play or the repro mode, there would've been no problem because the playing and the recording would be taking place at different heads.

Why not do all our bouncing in the play or the repro mode? Because the track we're bouncing to will not be in sync with the previously recorded tracks. If we had bounced our drummer's tracks onto Track 4 in the play or the repro mode, our mix on Track 4 would have been out of sync with Tracks 1, 2, and 3. Of course, once the bounce was done, we didn't need the first three tracks, so there really was no problem.

However, what about a situation such as this? We have recorded a voice on Track 1, lead guitar on Track 2, and bass guitar on Track 3. We want to combine the guitars (Tracks 2 and 3) onto a single track, so we bounce them to Track 4 in the play mode, to avoid feedback. The trouble is, when we play the guitars on Track 4 and the voice on Track 1, they're out of sync! There's a solution, though. Keeping the deck in the play or the repro mode, bounce Track 1 to Track 3. (Or to Track 2, if you prefer. Because we're not in the sync mode, there'll be no feedback.) Now Track 3 (voice) and Track 4 (guitars) will be synchronized.

External Bounce

Let's get our band back together. Our drummer has an attitude. This time, the drummer wants not only the three mikes we've set up on the left side, the right side, and the kick drum, but also another mike on the cymbal. This presents a problem. With the left side mike on Track 1, the right side mike on Track 2, the kick drum mike on Track 3, and the cymbal mike on Track 4, we have no track left to bounce to on our four-track deck. So we do the next best thing—an external bounce. We combine the tracks onto another tape recorder.

To do an external bounce (see Figure 11.28), you must have a second machine to bounce to. Another reel-to-reel deck is preferable, even one with only a single track. However, a cassette or cart machine will do in a pinch. Let's say we have a two-track deck in our studio, along with our four-track, and our four-track is filled with our drummer's tracks. We play the four tracks, adjusting the levels to taste, rewind the tape, and dub the mixture onto one of the tracks of our two-track. Bearing in mind that our goal is to put the percussion onto a single track of the four-track machine, we now have two options:

1. Dub our mix from the two-track onto a single track of our four-track.
2. Take the tape off the two-track (reels and all if you like), and put it onto the four-track deck.

Each of these choices has advantages and disadvantages. Option 1 requires us to dub from the four-track to the two-track, and back to the four-track, a loss of two generations. Remember that with each dubbing, the signal-to-noise ratio drops, due to the increase in noise. In this case, we know we're going to lose another generation when, after the other instruments and voices are put onto the four-track, we mix all our tracks down together. Our percussion track is going to add a lot of noise to our final tape.

Because of the increase in noise, I favor Option 2, in which we only lose a single generation. When the tape is manually transferred to the four-track, our percussion track will cover two of the four tracks. If we've bounced to the left track of the two-track deck, the recording will obviously take up half the tape width. When we move the tape back to the four-track, the recording will cover Tracks 1 and 2. Keep one of these tracks intact, and record new sound over the other. I'd advise keeping Track 2 because tracks at the edge of the tape (in this case, Tracks 1 and 4) tend to have more problems with noise and tend to have poorer signal quality when transferred from deck to deck than do tracks nearer the center of the tape. The only difficulty with this option is that the track you want to record over may not erase completely, due to differences in the head alignments between the two decks.

Bouncing constitutes one of the most important tools a producer has. If you're doing a project that's going to require a number of bounces, outline your strategy ahead of time. It's extremely frustrating (not to mention unprofessional) to be part way through a project and find yourself asking, "Now what do I do?" or "How do I get there from here?" Also keep in mind the price you pay for bouncing—noise. Do everything you can to minimize the noise. Use the fastest tape speeds and the widest tape tracks you can.

One more thing. Remember jumping tracks (from the editing chapter—Chapter 10)? If you're reading a lengthy narration, and you've jumped back and forth between Tracks 1 and 2 when you tripped, why not rewind the tape, and bounce the good takes on those two tracks onto Track 4 (or Track 3 if you're not in the sync mode)? It's really nice to be able to do all your work (recording, editing, mixing) on a single piece of tape.

DECK #1   DECK #2

Tracks 1, 2, 3 and 4 ◄─ ─ ─ ─┐   Track 1

                                              2

                                              3

                                              4

11.28 External bounce. Tracks from one deck are combined onto a single track of another deck.

Variable Speed Effects

Playing a tape back faster or slower than the speed at which it was recorded will make the pitch higher or lower than normal. You have a speed control on your tape deck that probably sets the movement of the tape at 7-½ or 15 ips. If you've ever recorded a track at 7-½ and played it back at 15, you know just how big a jump this is. Even if you were speaking v-e-r-y s-l-o-w-l-y at 7-½, the voice will not only double in speed, but also jump upward a full octave in pitch. It's very tough to double the speed and still maintain intelligibility!

However, many decks now have built-in oscillators, which allow you to alter the speed of the tape up or down, slightly or considerably, depending on the deck. Also, there are separate variable-speed devices available, which, when plugged into your deck, will allow you to slow your deck to a crawl, speed it up to double its top speed, or run at any speed within that rather considerable range.

If you don't have any electronic way to speed up or slow down your deck, try wrapping splicing tape around the capstan. This is one of the oldest tricks in the book, but it works. By wrapping tape around the capstan at the point where it contacts the recording tape, you're increasing the capstan's circumference. If your capstan is ½ inch in circumference, and turns at a rate of 10 rotations per second, then obviously when the pinch roller is engaged, the capstan will pull 5 inches of tape through in a second. If we were to increase the capstan's circumference to a full inch and keep its rate of rotation the same (10 per second), then we would now see 10 inches of tape being pulled through in a second.

Record a voice track, wind a few inches of splicing tape around the capstan, and, because the tape is being pulled through faster, the voice will rise in pitch.

To lower the pitch of the voice, record with the capstan already wrapped. Rewind and remove the splicing tape from around the capstan with a razor blade, and the voice pitch will be lowered. You're going to have to experiment with how much wrapping will cause how much pitch change. However, remember that a minor change in pitch will cause a major change in the character of the voice. Be very careful when you cut the splicing tape off the capstan. You know how important the capstan is to the tape transport system, so be gentle.

A lot of really creative production people do nothing more than combine different effects to produce new ones. For example, use your variable speed control along with your stacking or dovetailing. That's how you can re-create "The Chipmunks."

Record an instrumental track off a disc onto one track. Slow down the deck. Rewind the tape, and record your voice singing along with the music, but on a separate track. The music will of course sound slower and lower, but don't alter your voice. Rewind the tape, bring the deck back up to the original speed. The music is back to normal, and your voice sounds like a chipmunk.

Don't laugh. David Seville did rather well with this trick. He combined stacking (the three chipmunks), and variable speed dovetailing (the chipmunks and Dave talking to each other) for a clever and profitable effect.

By the way, who says that a sped-up or slowed-down voice has to remain that way? Why not gradually alter the pitch over time? In a spot for a tax preparation business, I had the voice get progressively lower, while talking about wading through the annual maze of tax forms. In contrast, in a spot about a vacation resort, which stressed the hotel's relaxing atmosphere and leisurely paced activities, I opened with a voice that grew progressively faster and higher pitched, bemoaning the rapid pace and stress of everyday life.

These effects can only be done if you have decks with built-in speed oscillators, or if you can plug a vari-speed into your deck. Again, easy does it. A little bit of speed alteration produces a lot of pitch change, and too much speed alteration makes for unintelligible speech.

Cross-fading

Cross-fading may be one of the most valuable weapons in your production arsenal. With it, you'll be able to take any piece of music, regardless of its length, and customize it to whatever length you want—without editing! You're going to need two decks, one of them a multitrack. Let's assume Deck 1 is multitrack and Deck 2 full-track or stereo. Let's also assume that we're customizing our music for a :30 spot, and that our voice track is :30 long and is already on the first track of Deck 1. You can use one method for cross-fading with music that is shorter than 30 seconds and another with music that is longer than 30 seconds.

Cross-fading Music Shorter Than 30 Seconds

When your music section is short, use the following method:

1. We'll assume that the music is only :20. Dub it from a disc onto Deck 2.
2. Backtime Deck 2 to the end of the voice track on Deck 1. Remember that when you *backtime*, you start elements of different lengths at different times so that they'll end together. You want the two tapes together because you need to know the exact point in the voice track (approximately :10 into the spot) to roll Deck 2. To find this point, cue the music on Deck 2 to the beginning, and the voice on Deck 1 to the end. Thread Deck 1 to roll backward, and hit both decks simultaneously. The music will roll forward, and the voice backward. The second you hear the music end, stop both decks. Rethread Deck 1 properly, and *note the word in the copy at which point the voice track has stopped.* This is your *key word*. From this point in the track to the end should be exactly the same length as the music track on Deck 2.
3. Recue the music on Deck 2, play both decks together, and see whether the tracks end together. They should.
4. Recue both decks. Put Deck 1 into the sync mode, and roll it. When the key word comes up, dub the music from Deck 2 onto another track of Deck 1. The two tracks should end together.
5. Recue Deck 1 (you don't need Deck 2 anymore), and turn down the pot to the music track.
6. Recue the disc.
7. Start up the disc and the voice track. When the key word comes up, the music track will be playing, but, because the pot is down, you won't hear it. You're going to do a cross-fade by potting the disc down while potting the music track up. The accompanying figure shows how your tape looks.

```
:30                :20                :10                :00
_____
* * * * * * * * * * * * * * * * * * * * * * * * * * * * * *  voice
_____
+ + + + + + + + + + + + + + + + + + + + + + + +  _____  music
              ! ! ! ! ! ! ! ! ! ! ! ! ! ! ! ! ! ! ! ! ! ! ! ! ! ! ! !  disc
```

Your voice tracks runs for the full :30, your recorded music is :20 long and you've backtimed it to start at :10 into the copy. And the disc, of course, is also :20 long. So, from :00 to :10 we hear voice and disc. From :10 to :20 we hear voice and either the disc or the backtimed music track. The cross-fade must take place within this time interval. After :20, the disc is gone, and before :10, the music track isn't available. Also, from :20 to :30, we hear both the voice and the recorded music track. You've jumped from the disc to the track, and the music fits the :30 of copy perfectly.

Cross-fading Using Music Longer Than 30 Seconds

What do you do when your music is too *long* for your spot? The following procedure shows you:

1. Let's say you have a 2-minute band on a disc that you want to customize to our :30 copy. Proceed as before, only this time, instead of dubbing the entire selection onto Deck 2, only dub the last 15–20 seconds of the cut onto the tape.
2. Backtime as before, and dub onto Deck 1 (in the sync mode of course) at the appropriate word in the copy.
3. Recue Deck 1 (voice and end-music) and disc.
4. Roll tape (music track potted down) and disc, and cross-fade from disc to music track. Because the music on the disc is longer than just 20 seconds, you have a lot more freedom. You can make the switch anytime after :10, as shown in the accompanying figure.

```
:30                :20                :10                :00

───────────────────────────────────────────────────
* * * * * * * * * * * * * * * * * * * * * * * * * * * * * * * * * * * * * * * * * *   voice
───────────────────────────────────────────────────
+ + + + + + + + + + + + + + + + + + + + + + + + +  _____   music

! ! ! ! ! ! ! ! ! ! ! ! ! ! ! ! ! ! ! ! ! ! ! ! ! ! ! ! ! ! ! ! ! ! ! ! ! ! ! ! ! ! ! ! ! ! ! ! ! ! ! ! ! ! !   disc
```

If the disc track is longer than 30 seconds, of course, you could forget the cross-fade and use the disc alone for the music, fading it out at :30. However, if the track has a solid, cold end (and I must admit I'm much more fond of cold ends than fades), the cross-fade will do the job.

Hints for Executing the Cross-fade

Executing the cross-fade isn't as easy as it looks. You don't just drop down one pot while cranking up another. The cross-fade is only effective if the listener can't hear it! It *must* be undetectable. Here are some hints:

1. Make sure the two pieces of music are in the same key. A longer piece of music may have one or more key changes, and consequently, may start and end in different keys. Cross-fading pieces of music in different keys is virtually impossible to hide.

2. Make the cross-fade under speech. Listen to the copy, and make the switch while the voice is holding the listener's attention. Don't cross-fade under a pause or during a break between sentences.

3. When executing the cross-fade, don't dump the disc and bring up the tape track at the same time. If you do, the overall level of the music will dip noticeably. Instead, slowly crank up the tape track until you can just hear it, leaving the disc pot alone. Then *quickly* dump the disc pot and bring the tape track pot up to the same level the disc had been. Very quickly! The cross-fade isn't a slow fade. The entire transition shouldn't take any more than a half-second. If the levels are consistent and the transition is under the voice, your cross-fade will never be detected.

Although the description of the cross-fade is lengthy, it's not nearly as difficult or as time-consuming as you might think. Once you understand the mechanics of the thing, exactly what you're trying to accomplish, you'll see how really valuable it is. If you have a four-track deck, it's even easier. Put your opening music on Track 1, voice on Track 2, and backtime the end music onto Track 3, synchronizing it with the voice on Track 2. All three tracks are on a single machine, so there's no need to juggle a deck and a turntable.

Additional Effects

You've probably noticed that all the tricks thus far involve reel-to-reel tape decks. That is not to say, of course, that you can't produce interesting effects on other pieces of equipment.

Wowing with Discs

Even though wowing is generally a production no-no, it has its place. Put enough discs on your turntable to hide the spindle (garbage records preferably because they'll probably get scratched). Put the disc you want to play on the top of the pile, with the hole off-center, and the music will sound to you the way it would to a guy who just scarfed down a six-pack.

Touch-Tone Telephone Jingles

If there's a touch-tone telephone in the studio, here's some information that might come in handy. The tones generated by our touch-tone phone for 1, 2, 3, 6, and 9 approximate do-re-mi-fa-and sol. Granted that's not much to work with, but you can still play a few tunes, such as "Jingle Bells" (3-3-3, 3-3-3, 3-9-1-2-3), "Mary Had a Little Lamb" (3-2-1-2-3-3-3, 2-2-2, 3-9-9, 3-2-1-2-3-3-3-3-2-2-3-2-1), Jack Benny's violin warm-up (1-3-9-6-3-6-2-3-1-3-9-6-3-6-2-3-1), "Shave-and-a-Haircut" (6-1-1-2-1, 3-6), and lots more. Also, you can create new (or beef up your old) jingles (see Chapter 14). Be careful not to inadvertently dial Singapore, and remember that after a few digits, the equipment at the phone company, which assumes you're dialing a real phone number, may cause bizarre sounds to come from your phone, so be prepared to record your tune a few notes at a time, hanging up the phone in between attempts. If your phone's wired directly into the board, great. If not, holding a mike by the earpiece of the handset will have to do.

Graphic Equalizer Special Effects

A graphic or parametric equalizer (EQ) can make a voice sound as if it's on the other end of the telephone, or coming through a cheap transistor radio (see Figure 11.29). Although telephone transmission is so much better now than it was just a few years ago, the old telephone sound is probably more useful to us because you'd probably want the phone voice to sound less than perfect, to contrast with the *proximate* (near) voice. The old phones had a frequency range of around 300–3,000 Hz, so on your EQ, droop anything outside this range, and boost 300–3,000 as high as possible. If your EQ is a parametric, set your low frequency to around 100 Hz and droop it, then set the high frequency to around 4,000 Hz, and crank it up. If you have a midrange control, set it wherever your ear tells you to do so. These are not hard and fast settings: always let your ear be your guide.

Old transistor radios had tiny speakers and, consequently, didn't reproduce low- and midrange frequencies very well. To simulate this, drop the lower half of your EQ down to the floor, raise the upper half to the ceiling, and run your voice track through it. You might also want to get rid of some of the upper midrange frequencies. Experimentation is everything.

Once, I wanted to simulate an old gramophone record, so I recorded the voice *dry* (i.e. free of any reverberation) and sent it through an EQ arrangement like this one. For added realism, I got a beat-up production record, put it on the turntable, and set the needle bobbing around in the end groove near the label. I sent this sound through the EQ along with the voice, and for good measure opened up a couple of unused pots on the console and cranked them as far up as I could to get some nice hissing. When I put everything together—voice, noisy disc, and hissing pots—and ran the whole thing through the EQ, it sounded just like an old Edison record.

11.29 Graphic EQ set up for telephone effect. (Photo: Courtesy UREI/JBL Professional.)

Summary

The point of this chapter isn't just to teach you a few special effects that'll sell your products and amaze your friends. The point is that most of these tricks can be generated with the simplest equipment. If you've got lots of gadgets around to play with, fine, but don't ever think that without these fancy electronic doo-dads your production has to be dull. Your only real limit should be your imagination, not your equipment. After all, who's in charge here? Don't be of the mind that you can go only as far as your machinery will allow: *you* decide what *you* want to accomplish, and make your equipment serve *you*.

Review

1. Look for atypical uses for your equipment. don't be limited by what the manufacturer says a machine can do.
2. The invisible edit can clean up the head of a tape recording by using the erase head as an independent tool.
3. A running start lets you pick up in the middle of a record or tape without any wowing.

4. Running the tape in the fast forward or the rewind mode across the heads (just for a second or two) produces a garbled sound, which can be used to indicate passage of time.

5. Nothing is useless. In production, everything has a place somewhere.

6. If your tape recorder doesn't have a built-in pitch change control, variable pitch change can be created with slight, varying pressure of the thumb on the edge of the supply reel during playback.

7. Tape loops stretch short pieces of music or sound effects to any length you like. With ambient sound loops, watch out for any punctuators that might stick out and be noticed with each passage of the loop.

8. Reverse threading makes the tape play backward. Make sure that your threading still keeps the tape against the heads.

9. Simple echo is created by recording with the deck in playback mode and with the deck's pot slightly open on the board. The faster the tape speed, the more rapid the echo. The more open the pot, the louder the echo will become.

10. A touch of echo can simulate reverberation, especially if the announcer is standing a short distance from the mike and speaking with increased volume.

11. *Reverse echo* puts the echo's decay on the front of the sound.

12. Despite the fact that most recorders can't play faster than 15 ips, by recording from deck to deck and adjusting speeds, an echo can be made to pulse at 30 ips, 60 ips, or more.

13. Closing the key switch on an echo that hasn't fully decayed results in a gated echo.

14. Using two decks, you can create a slap-back echo. Record a sound simultaneously on two decks, rewind them both to the beginning, and play them together but with one a fraction of a second behind the other.

15. Flanging is a two-deck effect in which the delay is only a few milliseconds, producing a weird, swishing sound. It's caused by the signals from the two decks partially canceling each other and shifting in and out of phase.

16. To cancel a particular frequency, the amount of delay needed equals 500 divided by the frequency.

17. A two-deck delay can be used to simulate stereo. Send the signal from one deck to the left and the signal from the other to the right. A short (⅛-inch) delay will cause the signals coming from the speakers to be different enough to produce a pseudo-stereo effect.

18. *Sweetening* is singing or speaking in unison with a prerecorded track.

19. *Chorusing* is like sweetening, but the process repeated twice or more.

20. *Slap-back echo* can be created on a single deck by recording a track, dubbing it onto another track of the same deck while in playback mode (bouncing), and then playing both tracks together.
21. *Multiple slap-back* is created by linking all the tracks of a deck together via patches, and having the signal imprint on each track in succession.
22. *Stacking* is like chorusing except that unison has been replaced with harmony.
23. *Dovetailing* involves alternating words or lines on different tracks.
24. An *internal bounce* combines tracks within the same machine. An *external bounce* combines tracks from one deck onto another.
25. Depending on the deck, bouncing from one track to an adjacent track in the sync mode may cause feedback.
26. Bouncing expands the number of tracks you can record on one deck.
27. When splicing tape is wrapped around the capstan, it increases the capstan's diameter, which can be used to create variable pitch change.
28. The cross-fade allows customization of a piece of music to any specific length.

12 ⬚⬚⬚
⬚⬚⬚
⬚⬚⬚

Stereo

The next major production area covered in this text is stereo—its theory and its technique. Stereo is what makes production jump from one dimension to two or from two to three, giving your work depth and breadth, and opening new creative avenues for reaching the audience. If your gear is already wired for stereo, fine. If not, don't worry. Even using standard mono equipment, you can still create wonderful, three-dimensional sound pictures.

Theory

Because you have two ears, what you hear—what your brain perceives as sound—is the sum of two separate and different signals, one signal from each ear. Because of their placement, your ears act like identical microphones with identical pickup patterns, pointed in opposite directions. The lines of information sent to the brain from each ear differ, just as the images transmitted by your eyes at any given time differ. It is this phenomenon, the fact that each eye has a slightly different perspective, that accounts for most of your ability to perceive visual depth.

The same holds true for your ears. If a pianist is playing off to your left, the amount of vibration reaching your left ear (because more of its pickup pattern is focused toward the source) will naturally be greater than the amount of vibration reaching your right ear. As a result, these two lines of information allow you to zero in on the location of the sound. Your head turns to the left until the amount of signal perceived by each ear is equal, indicating that the sound source is *equidistant* from each ear—that is, directly in front of your head.

Because your ears are on the sides of your head, they work best at locating sounds that exist in the same plane—left, right, front, and back. Up and down sometimes present a problem. Try this experiment. Blindfold a friend, snap your fingers once off to the left, and tell your friend to point to the source of the sound. Repeat this, having your friend pinpoint, as well as possible, snaps located in different areas (left, right, front, and back) around your friend's head. Finally, snap your fingers an inch or two under your friend's chin. This location may give your friend trouble. If successful, snap again under the chin, and see whether your friend picks a different direction. In any event, the fact that the sound was equidistant from both ears nullified the left–right homing ability of the ears. Because the snap under the chin wasn't within the plane of the ears' direct pickup pattern, location was even more difficult.

Keep in mind that the purpose of stereo is to re-create reality. A single speaker can do the job, as long as the original sound came from a single source, such as a single voice or a solo musical instrument. However, in the case of a complex sound, which had elements that emanated from a number of sources, such as an orchestra or a choir, the situation changes. When you hear an orchestra in concert, you hear different instruments coming from different parts of the stage—you don't merely hear the sum of all the instruments coming from stage center. A single speaker, while perfectly capable of reproducing the various frequencies, levels, and harmonics involved, can't convey the sense of the physical space in which the sound existed. Two speakers can. This is perhaps the most fascinating aspect of stereo: that you don't need a separate speaker for each element of the sound. Properly balanced, two speakers will suffice.

Image Placement

All of the the examples of stereo imaging that follow are based on two assumptions:

1. Your sound is playing through two speakers placed a distance apart. I've found that a minimum of 3½–4 feet of separation is necessary for good imaging. The distance between speakers determines the width of your auditory picture, the aural field. If the speakers are too close together, the signals spill over, and distinguishing left from right is difficult. Also, remember that you're creating aural images with width and depth. Placing the speakers too close together would be like watching a movie on a screen the size of a postage stamp. I also like to angle my speakers so that they face toward me, or at least in my general direction.

2. You are positioned equidistant from the speakers, like the apex of an isosceles (or better yet, an equilateral) triangle. Even if you're the apex of the triangle, if you're too far away from the speakers, the stereo effect will be lost. I don't know at what distance the loss of stereo happens, but I've found that if the distance from your position to an imaginary line drawn between the speakers is no more than double the distance between the speakers, you'll be okay. In other words, if a line connecting your speakers is 4 feet long, as long

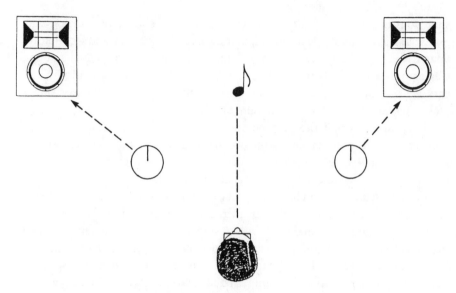

12.1 Identical signals playing at identical levels from both speakers will produce an image midway between the speakers.

as you're no more than 8 feet from the midpoint of that line, you'll hear the stereo without any problem.

Left–Right Placement

To examine stereo imaging, you need to first send a signal simultaneously into two pots, and then throw the output of one pot to the left speaker, and the output of the other pot to the right speaker. With mono equipment, do this. Patch a reference tone (1 kHz or so) into a multiple ("mult") on your patch panel (see Chapter 7). Then patch from the *mult out* into the left side of any pot on the console (pot A), and from the *mult out* into the right side of another pot (pot B). Adjust the pots so that their levels are identical. You are producing a mono image, and, if you close your eyes, you should perceive the tone to be directly in front of you (see Figure 12.1).

Using only the key switches, shut off pot B. Instantly, because the signal is coming only from the left speaker, your attention focuses full to the left (see Figure 12.2). Reopen pot B, and the image jumps back to the center. Shut off pot A, and the image jumps full to the right (see Figure 12.3). Thus, to position

a signal full left or full right, just make sure none of the signal is coming from the opposite side. Level is irrelevant. If you shut off pot B, regardless of the level from pot A, the tone will be perceived full left.

To position an image dead center, the signal has to emerge from both speakers equally. Again, the level is irrelevant. It doesn't matter whether the signal levels are high or low, just as long as they're equal.

Reposition the tone in the center with equal output from your two pots. Now, using pot B, slowly and continuously lower the level from the right speaker. As the level from the right side drops, the image will slowly move to the left until, when there's no signal coming from the right side, the image will be full left. Reversing the process will cause the image to shift to the right. As before, the overall levels don't matter. What's important is the *balance* of the left and right levels.

Try making the image move smoothly from full left to full right (see Figure 12.4). Start with pot B down (image full left). Gradually raise the level from pot B, and the image will slowly move toward the center, reaching the center position when pot B's level is the same as that of pot A. The instant the image reaches dead center, slowly fade pot A, and the image will move further and further right until, with pot A completely off, the image will be full right. If you have records in which sounds seem to move across your room from speaker to speaker, now you know how it's done.

On large multitrack consoles, the modules often come with a *pan control,* a single knob that places that channel's signal anywhere within the aural field. This control does what you've just done. It makes a duplicate of the channel's signal, throws one signal to the left and the other to the right, and then it automatically creates a balance that will position the image where you want it.

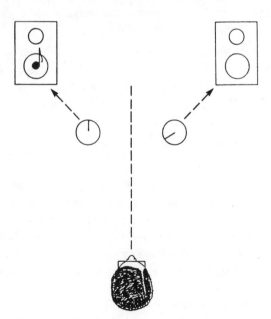

12.2 Image is full left.

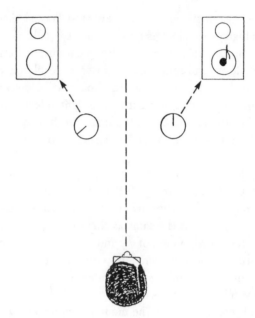

12.3 Image is full right.

Here's an important rule.

To position an image within the aural field (the space between the speakers), you'll need to use two pots, unless the image is to be placed full left or full right. (see Figures 12.5 and 12.6).

Remember, full left and full right only require a single pot (no balancing necessary). Also, although a center image requires an equal left-right balance, you can, using the mult, patch the signal into the left *and* right sides of a single pot. Turn the pot up, and the signal will come equally from both sides, producing a mono (center) image.

Front–Back Placement

Although stereo is primarily concerned with left–right placement, don't forget that your image can also have depth. Whereas with horizontal assignments overall levels weren't as important as the left–right balance, *depth* (or front-back placement, if you prefer) is achieved primarily by lowering the signal's level; however this alone won't create the illusion of distance. Two other factors need consideration: (1) reverberation, and (2) voice power.

Reverberation

If a voice (or any sound) is to be perceived at a distance, and if the sound is supposed to be within a closed space (for example, inside *any* size

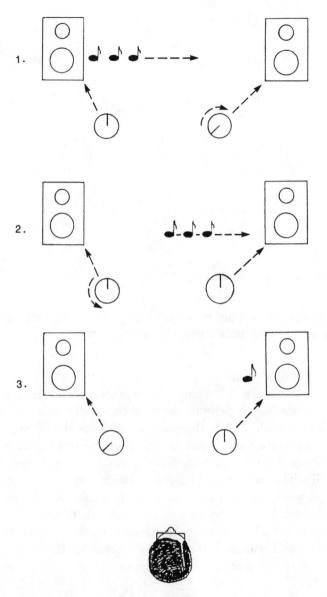

12.4 Moving the image from left to right.

room), the sound must reflect the reverberant qualities of that space, to sound realistic. If you want to show two people speaking, and one of the speakers is supposed to be significantly closer to the listener than the other, the distant voice should have significantly more reverb than the closer voice. If the voices or sounds are not supposed to be within a space, obviously no reverb is necessary.

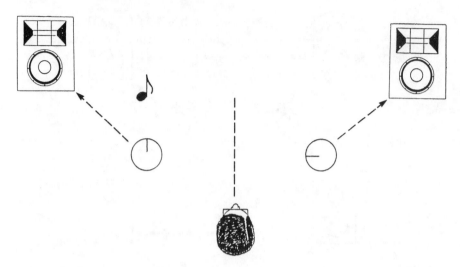

12.5 Left-of-center image placement.

Voices outdoors, for example, would not require any reverb because in reality, reverb is a phenomenon occurring within a closed space.

Voice Power

In Chapter 9, "Mixing," the point was made that a voice track must accurately reflect its environment. If a voice is meant to be distant, though the voice's level might be low, the power in the voice should be greater than normal. Dropping the volume isn't enough. Move off mike, or turn the mike away from you (if it's not an omnidirectional mike), or step back 10 feet or so. Speak the lines as you would if the situation you're depicting were real. Your meters won't read anything near 0 Vu, so crank the mike up until the voice is around –3. If the hiss becomes too noticeable (listen on your headphones), lower the mike pot a bit (unless you're planning to mix the voice with a noisy ambient track that will mask the hiss). Remember, as the voice gets further and further away, and as the level gets progressively lower, the power gradually increases.

Placement with Limited Pots

Because it takes two pots to position a sound somewhere in the field other than left, right, or center, it's not inconceivable (especially if you have a small or medium console) that you'll find yourself with fewer pots than you require to accomplish a particular stereo mix. Here's a tricky but effective solution.

Let's say we're recording a quartet, consisting of a soprano, an alto, a tenor, and a bass. In the aural field we want to position the soprano full left, the alto left of center, the tenor right of center, and the bass full right—just as they might appear on stage. This mix would require six pots, one each for the soprano and bass, and two each for the alto and tenor. Your studio is equipped with a four-track recorder, a two-track recorder, and a console with eight pots. However, on your console, four pots are wired for low-level signals (mikes and turntables), so you can't feed tape recorder signals into them. That leaves you with four usable pots—and you need six. Here's what you do.

1. Record the alto on Track 1, and the tenor on Track 2 of your four-track recorder.
2. Play back, and adjust the alto's level so that it's noticeably hotter than the tenor's. In the repro or the play mode (to avoid feedback), bounce this unbalanced mix onto Track 3.
3. Rewind, and again play Tracks 1 and 2 (keep Track 3 off for the time being), but this time make the tenor's level noticeably hotter than the alto's. Bounce this mix onto Track 4 (see Figure 12.7a). On Track 3, we have a lot of alto and a little tenor. On Track 4 we have the reverse—a lot of tenor and a little alto. If we throw Track 3 to the left and Track 4 to the right, the higher alto on the left with the lower alto level on the right will position the alto image left of center. Similarly, the high level of tenor on the right coupled with the low tenor level on the left will put the tenor image right of center. Got it? By combining the two voices in this fashion, we're able to position these two voices using only two pots instead of four.

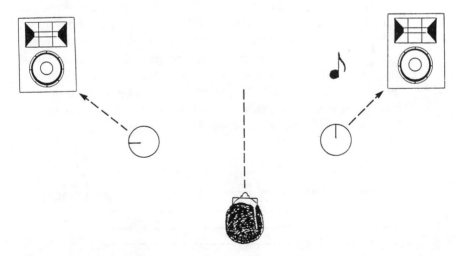

12.6 Right-of-center image placement.

4. Record the soprano on Track 1 and the bass on Track 2, in sync with Tracks 3 and 4.
5. Play all four tracks back, with Track 1 (soprano) to the left, track 2 (bass) to the right, track 3 (ALTO/tenor) to the left, and track 4 (alto/TENOR) to the right (see Figure 12.7b). Balance to taste.
6. If you want to mix these four tracks into a two-track stereo recording, roll the two-track in the record mode, recording the four tracks. Assuming that on the two-track, Track 1 is normaled to the left, and Track 2 to the right, our voice tracks will imprint with the four-track's Tracks 1 and 3 on Track 1 (left side) of the two-track, and the 4-track's Tracks 2 and 4 imprinting on Track 2 (right side) of the 2-track.
7. Play back the 2-track, and enjoy your quartet!

Here's another way of placing four voices across the field, using only four pots. I'm indebted to David Lanton, audio whiz, former student of mine, and all-around nice guy for this most creative solution to the problem. Incidentally, I'm sure Dave will be rich and famous some day, so remember that you heard about him here first.

1. Record one voice per track as you normally would. Let's say you've got the soprano on Track 1, the alto on Track 2, the tenor on Track 3, and the bass on Track 4.
2. Send Track 1 (soprano) to the left side of Pot 1, and Track 4 (bass) to the right side of Pot 4. The soprano is now full left and the bass full right.
3. Patch Track 2 (alto) into a mult, and bring two patches from the mult with one going into the *left* side of *Pot 2*, and the other into the *right* side of *Pot 3*. When you open Pot 2, the alto will come from the left speaker. When you open Pot 3, the alto will come from the right speaker.
4. Similarly, patch Track 3 (tenor) into another mult, and bring two patches out of the mult, plugging one into the *right* side of Pot 2, and the other into the *left* side of *Pot 3*. As it stands, Pot 2 brings up the alto on the left and the tenor on the right. Pot 3 does the opposite—alto right, tenor left. (Figure 12.8 shows all the needed patches.)

12.7a Track assignments.

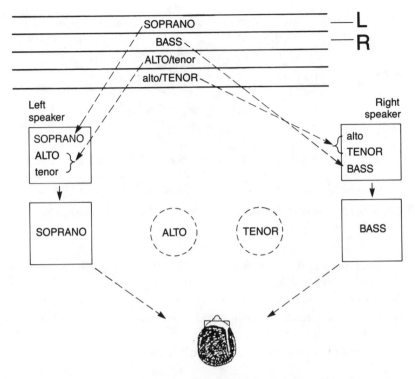

12.7b Final track assignments and image placement for stereo with limited pots.

5. Raise Pot 2 until the alto (left) and the tenor (right) are at a comfortable level. Then *slowly* open Pot 3. Two things will happen:
   a. As you raise the alto signal on the right via Pot 3, the alto image will move from the left toward the center. The more you open Pot 3, the further along the image will move until, when Pot 3 is the same level as Pot 2, the image will be dead center.
   b. Simultaneously, as the tenor signal from Pot 3 increases (left), it pulls the image sent to the right by Pot 2 toward the center.

I was really floored when Dave showed me this effect (which I call the *Lanton Effect*). With Pots 1, 2, and 4 open, he had the soprano and alto on the left, and the tenor and bass on the right. As he gradually opened Pot 3, the alto and tenor moved inward. Try it. You'll like it!

Comparing these two methods of placing images with a limited number of pots, Dave's is certainly easier than mine. It's a quicker method, too, because you don't have to do any bouncing. However, the alto and tenor are always equidistant from the center and are identical in volume unless you deliberately

12.8 Patching configuration for Dave Lanton's Effect. Solid lines are the alto patches, broken lines are the tenor patches.

record one at a lower level when you first lay down the tracks. My method allows for differences in volume, and, therefore, can place one of the two central images closer to the center than the other (your choice, depending on how you mix during the bounces). If you want equidistant placement of images, go with the Lanton Effect. Otherwise, try mine. Better yet, make up your own.

I love it when my students teach me stuff. Thanks, Dave.

Image Movement

Previously, we discussed how to make an image move within the aural field. Unless you have a board with a *pan* feature, moving an image laterally (side to side) always requires two pots, because the movement is the result of manipulating the speakers' signals independently. The overall level should remain constant, with the output of one pot rising as the other pot's level falls.

Although a sweep from full left to full right is quite dramatic, don't shortchange movement that covers less ground. Remember that movement (any movement!) will add realism to a recording.

Moving a voice or sound smoothly from front to back, or from back to front uses only a single pot for the sound itself (unless the sound is also moving laterally). If I want a voice to move to or from a distance, I'll record it while physically moving toward or away from the mike, all the while making the power in the voice reflect the distance from the mike. Then, all I have to do is play the track back, and add the concomitant reverb. If I'm giving depth and movement to SFX from a disc, I really can't control the power within the sound as I can with a live voice. I therefore have two options:

1. Record the SFX without alteration. Play it back, slowly and slightly fading the sound while similarly increasing the reverb.

2. Record the SFX beginning at full volume, and fade while recording. Play back without alteration, and add reverb.

The first method requires manipulating two pots (one for the track, one for the reverb), while the other requires moving only one pot (reverb), because the fading has been recorded on the track. In any event, the most effective results are often the product of *slight* volume shifts, and *slight* reverb change, so easy does it! When bringing a distant voice closer to the listener, gradually increase the voice's level while decreasing the voice power and the reverb.

Mixing in Stereo

The rules that govern mixing in mono still hold while mixing in stereo. However, a few additional rules arise. In general, if you're going to use stereo in commercial production, remember two things: symmetry and common sense.

Imagine that your aural field is on a balance scale, with the output of your left speaker on one side of the scale, counterbalancing the output of the right speaker on the other side. Our ear likes balance and symmetry, just as nature likes equilibrium. A voice coming from one speaker needs something coming from the other speaker to balance the production. The counterbalance doesn't have to be simultaneous. Two voices, one left and the other right, carrying on a conversation are symmetrical, even if the levels are different and the voices are talking at different times. Putting both voices on one side is not symmetrical. Putting the voice on the left and a music or ambient sound track on the right is not only asymmetrical, but it also doesn't make sense. These elements are supposed to blend, so why put them on opposite sides of the aural field? Just because you're able to create stereo, doesn't mean you have to use it all the time. Mono has its place. If you're mixing a solo voice and a music track except for the music, where's the benefit of stereo? Both these images (voice and music) should be in the center of the field. If you want to give the piece some breadth, run the music through the EQ (equalizer) for fake stereo feel. Further, there's absolutely nothing wrong or inadequate about leaving this particular production in mono.

If you're going to separate things within the aural field, there has to be a reason for your action. Above all, your production has to sound natural. If you hear a door knock on the left side, the sound of the door opening should also be on the left side. If the door opens on the left, doesn't it make sense that both voices, because they're both at the door, should also be on the left? Isn't this out of balance, everything being on the left side? Technically, yes but, as long as this imbalance isn't maintained for the duration of the spot, if the characters move toward the center or if one of them moves to the right side, you won't have a problem. A temporary imbalance is ok, as long as it's resolved.

In general, a music track should fill the field. The same holds true for ambient SFX. Having the entire aural field filled with *something* also helps mitigate any imbalances caused by the voice(s). Does this mean that music and ambient tracks should always be in stereo? Not necessarily. Let's say your spot has two people talking in a restaurant. Send the ambient sound equally out both speakers (mono), but separate the voices, one left and one right. Without the voices the background would exist only as a center image. However, *with* the addition of the *separated voices*, the *ambient sound* now *fills the space*. This is a remarkable phenomenon. The voices cause the background to stretch and fill the space.

Learning from the Greats

Mixing music in stereo is an art form in itself, requiring years of practice and experimentation. All producers have their own techniques and quirks, depending on the type of music being mixed. Producers of classical music most often seek to duplicate the arrangement and spread of the orchestra or ensemble they're recording. Popular music producers experiment more with sound placement (which is another reason why the sound a band makes on a record is often far different from the sound they'd make in a concert).

Most pop producers put most (if not all) of the percussion in the center of the aural field. The bass, too, is at or near the center. After that, anything goes, depending on the number and type of voices and instruments in the band. The best thing to do is to clap on a set of headphones and to listen carefully and analytically to the work of the great producers—people like George Martin, the innovative, self-effacing wizard who helped make the Beatles; Phil Spector, the first teen-entrepreneur and originator of the "wall of sound" in hits by groups like the Righteous Brothers, the Ronettes, and many others; Quincey Jones, the man who revived Michael Jackson's career with "Thriller"; Jerry Wexler and Lieber and Stoller, the guys who made Atlantic Records the wellspring of rhythm and blues, and who produced so many of the definitive hits from the early days of rock and roll for people like Ray Charles, Aretha Franklin, the Coasters, the Drifters, and others. If you want to learn music production, learn from the masters.

It's not that you can't learn from the current crop of producers. Listen to the work of guys like Chris Thomas (Wings, Pink Floyd, Elton John), Bill Szymczyk (The Who, Eagles, J. Geils Band), Roy Thomas Baker (Queen, Foreigner, The Cars), Richard Perry (Barbra Streisand, Carly Simon, Pointer Sisters) and others.

To my way of thinking, the most important producer ever to come down the road was the late John Hammond (1910-1987), the man who, over the course of over 40 years, recorded everybody from Bessie Smith and Benny Goodman to Bob Dylan and Bruce Springsteen, and in the process, showed how music and the music producer can be catalysts for social and cultural change. American pop music is enormously in his debt. He was our first great producer, and his incredibly eclectic taste and talent gave us classics in jazz, pop, folk, rock, and rhythm and blues. A social reformer and ardent integrationist all his life, Hammond ignored the unwritten rules of his day, and he assembled and recorded the first integrated band (the Benny Goodman trio with Teddy Wilson on piano in 1935). He put the first integrated band on a concert stage in 1936 (again led by Goodman), and spent over 30 years on the board of the NAACP. Early on, he recognized the merits of amplified instruments, and he was the first to put an electric guitarist (Charlie Christian) into a Big Band (Goodman's, naturally). The Newport Jazz Festival became integrated in the mid-50s because he told the promoter that a segregated jazz festival would be a fraud and a travesty. He was also instrumental (pardon the pun) in putting the first rock and roller (Chuck Berry) on the Newport stage.

The talented people he recorded form a veritable Who's Who of twentieth-century American music: blues singers Bessie Smith and Billie Holliday, jazz immortals Benny Goodman and Count Basie, bluesmen Sonny Terry and Big Bill Broonzey. Blues/gospel performers such as Joe Turner, whose work formed a chunk of the foundation of rock and roll. Folk singer/activist Pete Seeger, who Hammond recorded when all major labels blacklisted the singer (during the McCarthy era). Incidentally, do you know how "We Shall Overcome" became the anthem of the civil rights movement? The first time America heard it was when John Hammond recorded Pete performing the tune at his Carnegie Hall concert. The record became a hit, and the song became an icon. Aretha Franklin, George Benson, Bob Dylan, Bruce Springsteen, and poet Allan Ginsberg all found their way to John Hammond, or he went looking for them. You can't listen to contemporary music and not hear the influence of this singular man. He produced over 100 albums for us, and yet, sadly, few outside our profession know of his name, of his accomplishments, or of his importance.

Never lose sight of the fact that stereo, like most of the things in this book, is a tool, a means of achieving an end, and not an end unto itself. It's easy to overlook this fact and to go crazy with stereo separation, because it's fun and exciting. However, you should start from the end. What do you want to accomplish? Will stereo help? Will it add something that mono won't? Is it worth taking the extra time to create stereo? As you plan a mix, don't automatically plan on stereo without asking yourself these questions. Then, as the knight says in *Indiana Jones and the Last Crusade,* "Choose wisely."

Review

1. Stereo creates multidimensional images by sending different signals to the speakers.

2. A signal emerging equally from both speakers will be perceived as being midway between them.

3. When the amount of signal emerging from the speakers isn't identical, the image will be perceived as being closer to the speaker with the louder signal.

4. By balancing each speaker, an image can be positioned anywhere between them or can be made to move within the field.

5. Reverb and vocal power are used to position an image closer to or more distant from the listener.

6. On a console without panning capability, generally, you'll need two pots to position an image anywhere other than full left, full right or dead center. However, there are ways to position images using a limited number of pots.

7. Moving an image laterally involves continually and smoothly varying the individual levels coming from the speakers. Front-to-back movement involves varying the overall power level and the amount of reverb.

8. When mixing in stereo, keep the field symmetrical and logical.

9. A music background should almost always fill the field.

13

Processing

I must confess a deep fascination for gadgets and gizmos of all kinds. In the world of production, we have more than our share of electronic toys. Some are of limited use and, as such, are only rarely found in radio and television production studios because budget considerations tend to rule out the purchase of such tools, which are frequently expensive. On the other hand, some devices are not only commonplace but are also integral to the production professional's arsenal. I've lumped these diverse devices under the common heading of *processors*, indicating that, even though they're all different, they all alter an input signal in some fashion.

Equalizers

On your stereo system or amplifier, do you have a switch, knob, or slide of some sort labeled *Tone* or *Bass/Treble Control?* If so, then you've already had some introduction to equalization. An *equalizer* (EQ) is simply an amplifier, just like the amplifier that controls your stereo's volume—with one distinction: whereas your stereo amp raises and lowers the level of the entire signal being fed through it, an EQ allows you to raise (boost) and lower (attenuate, droop, cut) the volume of *part or parts* of the signal, while letting the rest of the signal pass through untouched.

The EQ breaks the signal into individual frequencies or groups of frequencies, called frequency bands, and the operator is able to alter the dB level of one or more of these frequencies or bands.

Although EQ is often used in commercial production, its most critical use is in music production. By increasing or lowering the levels of various parts of the signal, a producer can totally change the color of a musical sound. Here's an arbitrary listing of frequency ranges and which instruments or characteristics of the music they affect.

lower bass (10–80 Hz): rumble, richness, depth—boosting this area will also boost line hum (60 Hz).

upper bass (80–200 Hz): power, energy (especially percussion and bass) *lower midrange* (200–500 Hz): warmth in vocals and strings (also backup instruments like guitar, keyboards)

mid midrange (0.5–2.5 kHz): sharpness of lead vocals, lead instruments, and some percussion (snares)

upper midrange (2.5–5 kHz): brightness, presence— harmonics located are within this and higher ranges

lower treble (5–10 kHz): sibilance, sharpness of cymbals—location of tape hiss

upper treble (10–20 kHz): inaudible to most people (in the upper end of this range), simply cut out by some studios, but I don't think this is such a good idea because important harmonics exist here (You may not be able to hear them distinctly, but they add to a sound's timbre; people with sensitive ears claim this range adds delicacy to high-frequency tones.)

The amount of EQ control you have over a signal depends on the type of equalizer you're using. Most of today's equalizers fall into two categories— parametric and graphic.

Parametric Equalizers

A parametric EQ system is the type built into multichannel console modules. It usually consists of at least three knobs: one for selecting low frequencies, one for midrange, and one for treble. Each knob can be set at one of a number of frequencies (usually four or more) within its particular range. The frequency selector knob may have fixed positions (click-stops) allowing you to choose only from a few preset frequencies, or it may be a sweep control, letting you set the knob anywhere within the control's range.

You can sometimes select two frequencies in a given range using a parametric EQ because the frequency controls probably have overlapping ranges. A bass control might cover 50–500 Hz, midrange 0.2–5 kHz, and a treble 2–15 kHz. As a result, you could control two lower frequencies—for example, 70 Hz and 200 Hz—by selecting 70 Hz on the bass control and 200 Hz on the midrange control.

Concentric to or adjacent to each selection knob is a volume control, allowing you to raise and lower the dB level of the selected frequency 5–10 dB or more.

Finally, there may be a bandwidth control. When you raise or lower a selected frequency, you also raise or lower frequencies on either side of it, although to a lesser degree. Imagine plucking the middle of a taut guitar string. The string's greatest displacement is at the point where your finger pulls it off center. However, there is also displacement on either ide of the center point, and the amount of displacement decreases as the distance from the center increases.

The bandwidth control determines how far to either side of a selected frequency the equalization is to be felt. The control itself is generally a knob that sweeps from *narrow* to *broad*, a *narrow band* being only a few Hz on either side of the selected EQ frequency, and a *broad band* covering possibly the entire frequency range of the control. If your EQ doesn't have a bandwidth control, then the bandwidth is fixed by the manufacturer.

Graphic Equalizers

A graphic EQ uses a series of slides to select frequencies for equalization. Each slide lets you raise or lower a frequency band 12 dB or more, and each slide covers a fixed band of frequencies: there is no bandwidth control on graphic EQs. Whereas a parametric EQ only allows you to choose a single frequency in a particular range (bass, midrange, treble) at one time, the graphic EQ allows you to affect as many frequencies as there are slides.

To be of any real use to you, a graphic EQ should cover the entire sound spectrum in no more than one-octave intervals. If the frequency of a slide is twice the frequency of the previous slide, then you have a one-octave EQ. (When you raise a tone by an octave, you double its frequency.) A one-octave EQ may have a frequency array like this (in Hz):

62.5, 125, 250, 500, 1,000, 2,000, 4,000, 8,000, 16,000

Each slide has a frequency that's double the previous slide, therefore one octave higher. This arrangement would allow you to control any or all of nine frequency bands at one time. Little three- or four-band graphic EQs, such as the ones associated with car stereos and inexpensive home gear, are useless in studio production because the bands are much too wide to be of any value. Using this sort of an EQ would be like shooting mosquitoes with a cannon.

The big disadvantage of a graphic EQ is that it can affect maximum equalization only on the frequencies assigned to the slides. If you have an EQ like the aforementioned one, and you want to boost the 700 Hz component of a signal, you have a problem, because you have a slide at 500 Hz and one at 1 kHz (1,000 Hz), but nothing at 700 Hz. All you can do is boost the slide that's closest to your target frequency, in this case 500 Hz, and hope for the best. As a result, I find a graphic EQ fine for overall coloring of a sound, but if I want to really fine tune a signal and bring out or drop down specific voices, instruments, or other sonic components, I want the precision of a parametric EQ.

If you have to use a graphic EQ, try to get yourself a ⅓-octave model (see Figure 13.1). It'll cover the same range (around 40 Hz to 15 kHz), but in ⅓-

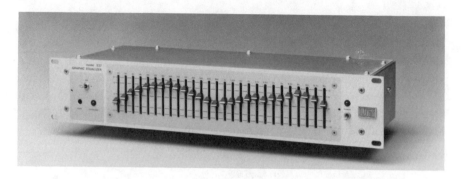

13.1 A one-channel 1/3-octave graphic EQ. (Photo courtesy UREI/JBL Professional.)

octave intervals, so you'll have anywhere from 27 to 30 slides, giving you three times the flexibility of a 1-octave EQ.

There are graphic EQs that are switchable from stereo to mono. The slides are grouped into two 1-octave arrays, one for the left side and one for the right side. However with the flip of a switch, the frequencies of the slides are altered so that the two arrays function as a single ⅓-octave mono EQ. This is a great tool!

Although equalization is crucial to music production, it's also useful in commercial production. If you have a tape with too much hiss, run the signal through your EQ and droóp the frequencies over 8 kHz. If a tape's muddy because of too much bass, droop everything below 100 Hz. If you hear line hum, cut the signal at 60 Hz. In commercial production the EQ is often used as a strictly cosmetic device, something to fix, or at least improve, defective recordings. Also, of course don't forget that the EQ can be used to generate a few special effects (see Chapter 11).

There is no perfect equalizer. Every model has its own advantages and disadvantages, in terms of frequency selection (variety and quantity), bandwidth, miscellaneous features like filters (more about these later in this chapter), and stereo/mono switchability. The EQ you buy depends on the application you have in mind.

More Fake Stereo

In the special effects chapter (Chapter 11, "Tricks of the Trade") we simulated stereo by sending the same signal out both speakers, but delaying one side a few milliseconds behind the other? The result was that the waveforms emerging from the two speakers were not exactly identical, thereby creating the illusion of stereo. Here's another (better) way to generate fake stereo, using equalization instead of delay. Although I suppose a parametric EQ could be used to simulate stereo, a graphic EQ is my choice. You should also have a four-track deck. Here's what you do:

1. If you have a mono EQ (a single set of slides), first record your sound onto Track 1 of your four-track recorder.
2. Adjust your EQ so that the odd-numbered slides are raised to the maximum, and the even-numbered slides are dropped to the bottom.
3. With the deck in the repro or the play mode, send your track through the EQ, and record this equalized signal onto Track 3.
4. Rewind the tape, and readjust the equalizer the opposite way (odd slides down, even slides up).
5. Send Track 1 through this new equalization arrangement, and record onto Track 4. What you've done is to record the same sound onto Tracks 3 and 4, but with opposite equalization patterns.
6. Send Track 3 to the left and Track 4 to the right, and play back. If you don't want to tie up two pots, send these two tracks into the left and right sides of a single pot. The effect is the same.

Individually, Tracks 3 and 4 don't sound too good; in fact they probably sound awful. Don't worry. When they're combined, they even each other out. The result is so even that you may not immediately notice the stereo effect. If you close your eyes, though, you'll hear that the overall sound isn't positioned in the center—the sign of mono.

If you happen to have a stereo EQ, you can save a couple of steps (see Figure 13.2).

1. Send your original signal into a mult, and bring two patches from the mult, one into the left side of the EQ and the other into the right side.
2. Adjust the left side of the EQ with the odd slides up and the even slides down. Do the reverse with the right side.
3. Roll the tape in the repro or play mode (sending the signal to the mult, where it's split and sent into both sides of the EQ), and record the output of the EQ (left and right sides) onto two available tape tracks. Throw one of these tracks to the left and the other to the right, and play back.

I like this kind of fake stereo better than the method using delay. It's easier and it sounds more like stereo which is, after all, what we're trying to accomplish. Don't let the term *fake* stereo turn you off. There's nothing cheap or sleazy about it. Assuming that your station broadcasts in stereo, there are a number of instances where you'd use fake stereo. For example,

13.2 Both sides of a stereo EQ set for fake stereo. (Photo: Courtesy UREI/JBL Professional.)

1. Commercials from outside agencies may come to you produced in mono. As you dub to cart, first run your mono spot through a two--channel EQ arranged for fake stereo.
2. Oldies from the 1950s and early 1960s, which were originally produced in mono, can be recast with a judicious application of stereo. Be careful, though. It's not a good idea to radically alter something your audience considers classic.

Compressors and Expanders

In a nutshell, a *compressor* (often referred to as an AGC—automatic gain control) is designed to control a signal's *dynamic range*—that is, the decibel difference between the loudest and the quietest sounds in a recording (see Figure 13.3). The compressor does so by means of a *compression ratio*—that is, a ratio of the signal level going into the compressor versus the level of the signal coming out. For example, a 2:1 ratio would mean that if we increased the level of a signal going into the compressor by 2 dB, the signal emerging from the compressor will have increased only 1 dB. For every 2 dB of change going in, 1 dB of change comes out. If the signal going into the compressor

13.3 Compressor/limiter. (Photo: Courtesy UREI/JBL Professional.)

jumps 10 dB, the signal coming from the compressor increases only 5 dB. In this way, a signal with an exceptionally wide dynamic range will have that range compressed. This principle is important if the signal's dynamic range is greater than the tape can handle. Without compression, you'd risk saturation and, consequently, distortion.

The same holds true for low-level signals, but in reverse. When the signal level gets too low, the compressor, according to the compression ratio, lowers the volume of the output signal less than the change in the input signal. So at a 2:1 ratio, if an already low level signal going into the compressor drops by 6 dB, the compressor only allows the signal coming out to drop by 3 dB.

This aspect of compression is important if a solo voice or instrument within a group has a drastically lower level than the other voices or instruments. My first digital album was a recording of Pachelbel's *Canon in D*, a beautiful piece that was used in the movie *Ordinary People* and in a number of commercials, most notably in a spot for G.E. light bulbs. I'm sure you'd recognize it if you heard it. On my record, there's a harpsichord solo that is virtually inaudible and would probably go unnoticed by anyone not wearing headphones. What was needed at the recording session was either a separate mike on the harpsichord or compression of the harpsichord's signal. Aficionados and producers of classical music don't like their music tampered with. Fair enough, but I can't see the advantage of being a purist if the sound can't be heard.

Depending on your compressor, you might be able to select a particular ratio from a number of available ratios, such as 2:1, 3:1, 5:1, 10:1, and infinity:1. Infinite compression would mean that no matter how the level of the input signal changes, the signal coming out of the compressor wouldn't change much at all. Everything would sound the same—loud passages, soft passages, everything. Talk about dull!

Your machine may also have a *threshold* control. This allows you to select the dB level at which the compression will begin. A signal that has a level below the threshold will pass through the compressor unaltered. When the signal's level goes above the threshold, the machine begins compressing the signal according to the selected ratio.

An *expander* performs the same function, but in reverse. Ratios are along the lines of 1:2, 1:3, 1:5, 1:10. For every dB of change going in, 2, 3, 5, or 10 dB of change come out. Often, the compressor and the expander will be combined into a single device termed a *compander*, which can perform both functions on a signal simultaneously.

You may also have attack time and release time controls on your compressor. The *attack time* allows you to control how fast the machine reacts to incoming signals, and the *release time* allows you to affect the time it takes for a compressed signal to return to normal. These controls are calibrated in milliseconds, and their effect on a signal can be gauged only through practice and individual taste. Depending on what I'm recording, I tend to favor a fairly short attack time, especially when I'm producing something with a lot of punch (i.e., sharp signals with very short duration, such as punctuators or fast music). If the attack time is too short, musical notes (and percussion) will have their power cut. If the attack time is too long, the machine will make a mess out of musical passages with lots of rapid sounds.

I also like a fairly long release time because it tends to smooth out a track. If the release time is too short, and you're using a high compression ratio, when silence of any duration hits the compressor (a pause between sentences, a rest in the music, etc.), the machine sends out a pumping, hissing noise. A friend of mine once compared this sound to the sound of a balloon being inflated. If the release time is too long, the signal, especially if it's fast speech or music, will be muddy, as released signals will run into newly attacked signals.

If you want to hear something weird, set your compressor for the shortest attack and the longest release times you can. A signal run through this setup, especially with a high compression ratio, will sound like it's playing backward.

Experimentation is the key to success with compression and expansion because there's no pat formula that will work for all occasions. In the 1960s, a lot of the country's top rock and roll radio stations were compressing the daylights out of their broadcast signals and adding a ton of reverb. Times change, fortunately, and nowadays, a more natural sound is desired. In general, compression should make the elements in a produced piece sound more unified. If your work sounds okay without compression, don't bother with it.

Noise-Reduction Systems

Compression and expansion are also used in Dolby and dbx, the most popular noise-reduction systems around today. For example, let's say we're recording a musical group, and they're playing a piece with an exceptionally wide dynamic range. If there'd be any place where noise would be a problem, it would be, of course, in the quiet passages. By recording with compression, when the band plays those quiet sections, the compressor will keep the low level signals from getting too low, thus somewhat masking the noise. I hear you ask, " But won't the compressor also lower the volume of the loud sections, and reduce the dynamic range of the piece?" Yes, but if we then run our compressed signal through an expander, two things happen:

1. The original dynamic range is restored.
2. Because noise is low level, the expander pushes the noise level even lower.

Expansion will push the noise level low enough so that it will be virtually inaudible. Remember, we're talking about noise *reduction*, not noise elimination. If there's noise present, but you can't hear it, you've won the battle. The noise-reduction systems we use are called companders because they utilize this compression/expansion method to reduce noise.

Dolby

The audiocassette industry owes a whopping debt of gratitude to Dr. Ray Dolby. When cassettes came onto the market in 1964, they were awful. Sure they were compact and easy to store and use, but because of the narrow tracks and the slow tape speed (1-⅞ ips), the sound was chock full of noise. Except for the novelty value, they probably would've gone the way of the dodo—if it hadn't been for Dr. Dolby.

Dolby developed three systems, and labeled them A, B, and C. Dolby A is for use in professional recording studios. Dolby B and C are designed for smaller studio and, to a degree, home use. Your cassette deck probably has Dolby B, if not both B and C built right in.

Dolby A splits a signal into four bands (80 Hz and below, 80–3,000 Hz, 3–9 kHz, and above 9 kHz), and it applies separate compression/expansion to each band as needed—that is, when the signal's level gets too low. Each band contains a different type of noise. Rumble and line hum are found in the lowest band, midrange noise and channel crosstalk exist in the low midrange band, and tape hiss is found in the upper midrange and treble bands. Any noise between 50 Hz and 5 kHz is reduced 10 dB, increasing to a 15 dB reduction at 15 kHz.

Dolby B was intended for nonprofessional use. Noise reduction is applied only to a single high-frequency band, from 10 kHz on up. The noise reduction is a flat 10 dB across this band.

Dolby C, like Dolby B, acts on a single high-frequency band, but with 20 dB of noise reduction.

Dolby is a great system, but remember that it's designed to be cosmetic, to fix a signal that has a noise problem. Don't use Dolby unless you have to.

13.4 A dbx Type I noise reduction system. (Courtesy of AKG Acoustics, Inc.)

Here's why. If your deck has Dolby C, keep the Dolby off, and play a good-quality cassette. While the tape is playing, switch the deck into Dolby C mode. Instantly, the upper end will sound like somebody threw a blanket over it. Yes, the noise is cut way down, but so is the harmonic content and the brightness of the signal.

dbx

The dbx system differs from the Dolby system in a couple of important respects. First, dbx works on the entire signal at one time, rather than breaking the signal into separate frequency bands. Second, dbx uses a flat 2:1 compression ratio on the signal, and then a 1:2 expansion ratio, which reconstructs the original dynamic range and squashes the noise down 30–40 dB or more.

There are currently two dbx systems, dbx I and dbx II: Type I is for studio use (see Figure 13.4), and Type II, which is less critical, is for broadcast and, to a lesser extent, home use.

In the debate over the relative merits of Dolby versus dbx, both sides have their adherents. Dbx is more sensitive and offers greater reduction than Dolby, so it's the system of choice for audiophiles, who tolerate noise about as much as a dentist tolerates plaque. If you check your album and CD covers, you'll find that the overwhelming majority of them will indicate use of dbx noise reduction. Dolby is not as sensitive as dbx, which makes it perfect for film sound. After all, in the average theater the environment contains a lot more commotion than your home listening room, what with the kids in the front row throwing popcorn at each other, and the couple seated behind you exploring the depths of passion.

Reverberation

As has been previously mentioned, reverb makes a track sound full and natural by giving it the same coloring as its environment. Also, as you'll remember from Chapter 11, putting an identical touch of reverb on different sounds helps them to blend. Reverb is generated in four ways: a reverb chamber, plate reverb, spring reverb, and digital reverb.

Reverb Chamber

A *reverb chamber*, as its name implies, is a room with hard, reflective surfaces. Within its confines are placed a speaker for playing the sound, and one or more mikes, positioned to pick up the sound as it bounces around the room. In terms of natural-sounding reverb, the reverb chamber is the best. After all, it *is* natural. Different reverb times are created by removing or repositioning movable sound absorbers called *baffles*. The room must be large enough to accommodate bass frequencies with large wavelengths. Your bathroom, despite the great reverb it gives your voice when you're in the shower, won't do. It's too small and can't give you the wide range of delay times a big room can. The room can never be too big because you can always cut down its acoustic size with baffles. Your chamber should be at least 12 × 12 ×12 feet (1,728 cubic feet). The only drawbacks of the reverb chamber are (1) the expense, and (2) the time required to move the baffles around to get the reverb you want.

Plate Reverb

A reverb plate is a large steel sheet the length and width of an average door. The plate is suspended on its side in a wooden frame, and is outfitted with a couple of electronic devices: (1) a driver, which acts like a speaker, transmitting a *dry* (free of any reverb) signal to the plate, thus making the plate vibrate; and (2) a small mike, which transmits this mixture of dry signal and plate vibration back to the console. The sound from a plate reverb is pretty good, but the plate isn't cheap, and it's very cumbersome. Also, loud signals tend to become muddy.

Spring Reverb

The spring reverb operates similarly to the plate reverb, but instead of making the dry signal cause a plate to vibrate, the signal is sent into an array of stretched springs. These springs differ in tension and density, and when they're set to vibrating, the noise they generate, when combined with the dry signal, produces something akin to reverb.

There's good news and bad news about spring reverbs. They're relatively inexpensive (as opposed to large plates or reverb chambers), and they're easy to use. They are small (component size), enough to fit right into a rack along with your compressor and your EQ. Many allow you to vary the decay time and the equalization, thus allowing you to simulate more environments. However, cheap spring reverbs turn out a twangy, unnatural sound that smacks of something from the Twilight Zone. I used a spring reverb for many years, which worked fine when the reverb level was low. However, when I cranked it up to simulate a really large room, it'd produce *boings* and *pings* that would drive me crazy. Like most other production tools, if you're going to get a spring reverb, spend the extra bucks, and get a good one.

Digital Reverb

It's ironic that the reverb system that is the most totally electronic produces the most natural-sounding reverb outside of a reverb chamber. Here, the signal from the console is sent into the reverb unit and repeated many times per second at varying intervals of time (milliseconds apart), with a gradually decreasing amplitude. The result is a perfectly natural sounding reverb that can be used to simulate the acoustical environment of virtually any enclosure. Many digital reverbs come with a number of presets for your convenience, allowing you to re-create particular environments with the push of a button. If you think that digital reverb units are expensive, you're right. If you think that digital reverb is my choice, you're right again.

For a lot of producers, whether commercial or music, their use of reverb, EQ, and compression makes their work as readily identifiable as does an artist's signature in the corner of a painting. As you experiment, progress in your mastery of technique, and become more adept at handling these critical production elements, you'll develop your own style. Above all, that's what this book is designed to help you do.

Other Processors

Outside of compressors, reverbs, and EQs, which are mainstays of the production studio, there are other gadgets you may encounter, which run the gamut from the useful to the frivolous. Here are a few of them: varispeeds, limiters, filters, pitch shifters, digital delays, de-essers.

Varispeed

If your tape decks don't come with variable speed controls to make the tape run faster or slower, you can buy a varispeed that plugs into your deck's motor and allows you to slow the capstan to half-speed (or slower) or to double the capstan speed. Besides this wide range, the varispeed also allows you to set the speed anywhere within that range, not merely at a few positions preselected by the manufacturer. On the negative side, the varispeed can put a real load on your deck's motor, especially if the motor is old. After using the varispeed, shut it off.

Limiter

A *limiter* is similar to a compressor, in that it controls amplitude. However, unlike the compressor, which diminishes the amount of gain through a compression ratio, the limiter actually sets a loudness ceiling that a signal cannot exceed. No matter how loud the signal gets, the limiter allows it to reach the set-point and no further. Unlike compressors, limiters have no effect on low-level signals. You may not run into a limiter in your studio, but radio stations use them in their broadcast chains to ensure that the signal they broadcast stays within the parameters set forth in the station license.

Filters

A filter works like an EQ, in that when a signal is input, some part of it is affected while the rest of the signal is untouched. However, whereas the EQ merely boosts or attenuates frequencies, a filter removes part of the signal almost completely. Also unlike the EQ, you have no control over a filter's operation, other than turning it on or off. Filters are usually built into other components like console modules or equalizers. Here are the most common types:

1. High-pass (or low-cut) filter— removes the frequencies below a preset point. Good for eliminating rumble and line hum.
2. Low-pass (or high-cut) filter—removes frequencies above a preset point. A substitute for noise reduction, it can be used to cut out tape hiss and other tape or record noise above 8–10 kHz, but it'll also cut out the harmonics of your recording.
3. Mid-pass (or band-pass) filter—sets low- and high-frequency cutoff points and eliminates everything outside them, thus allowing only frequencies between these limits to pass through.
4. Notch filter—cuts out a single frequency, or an extremely narrow band of frequencies. It's better than a low-cut filter in removing line hum (60 Hz), because it removes only the offending frequency, instead of lopping off everything from 60 Hz on down.
4. Comb filter—works like a series of notch filters. Flanging (see Chapter 11), gives the effect of a comb filter because it's a series of canceled frequencies.

Pitch Shifter

To alter pitch generally requires an alteration of speed. Not so with the digital pitch shifter, which allows you to electronically alter pitch without affecting time. This capability is especially handy if, say a :32 spot comes into your studio. You speed up the tape any way you want so that it runs a flat :30. This will naturally cause the pitch to rise. You then send this signal through the pitch changer and drop the pitch back to normal. Your spot sounds fine, and it's :30.

 Pitch shifters are great fun and can produce a number of effects by letting you shift the pitch an octave or more up or down. Beware of glitches, though. If you vary the pitch too much, you're apt to hear sporadic snaps, crackles, and pops on your recording.

Digital Delay

A *digital delay unit* (and there are lots of them on the market, from cheap plastic gizmos to units costing thousands of dollars) takes an input signal, digitally duplicates it, and then sends both signals out, with one delayed by as little as a few milliseconds, or by many seconds. With a delay unit, you can create many of the echo effects from chapter 11 (and a slew of other effects) much more quickly and easily than by using only tape recorders. Delay also provides a 7 to 10-second safety net for radio talk show hosts, who can eliminate objectionable language before it reaches the air.

If you couple pitch shifting and delay you can produce a wealth of effects, from the practical to the playful. If you're in the market for such things, try to get a machine that combines both pitch shifting and delay capabilities in a single component. The Harmonizer® by Eventide Clockworks comes to mind.

De-esser

Some people have a special problem when pronouncing words containing the letter *s*. Their *s* sounds raspy, overly sibilant, and sometimes downright soggy. Pinky Lee (a cherished part of my childhood) had this problem. So does Donna Dixon, the statuesque blonde from the TV show "Bosom Buddies." Enter the de-esser, a device that, depending on the manufacturer, employs a filter and/or a limiter to reduce sibilance problems within the upper midrange where these annoying sounds are produced.

New Technology

Every time I pick up one of the broadcasting or production trade magazines and check the ads throughout, I'm amazed at the array of processors available. As our technology continues to leap forward, the number of tools and toys increases. If you have a limited budget (and who doesn't?), and you're involved primarily with commercial radio production, first buy a good EQ (parametric, with sweepable frequency selection and as many features as possible) and a good reverb (digital or top-of-the-line spring). These are critical tools. Then go for a compressor/expander. If you're a music producer, you must have these three units plus a noise-reduction system, either Dolby A or dbx I. These components represent the *minimum* processing capability you need to have on hand. Special effects generators and fancy filters can wait until you've got some money coming in.

Finally, a word of advice. Beware of the producer's *kiss of death*—the rut. It's easy to find a nice EQ or compression setting and use it on everything you turn out for the rest of your life. Don't! No single mode on any piece of equipment is going to fit all the types of production you're bound to encounter. Creativity, by its very definition, implies variation. Don't get too comfortable. Comfort leads to repetition, which leads to ruts. Stay clear of ruts! Your competition needs a place to live.

Review

1. A *processor* takes a signal and alters it in some fashion.
2. An *equalizer*, parametric or graphic, breaks a signal into frequency bands and allows you to raise or lower the level of these bands without affecting the rest of the signal.
3. Equalization is used to color a sound, especially music, and to bring out such characteristics like warmth, brightness, power, etc.
4. A *compressor* restricts a sound's dynamic range.
5. An *expander* does the opposite of a compressor, stretching a signal's dynamic range.
6. Noise reduction combines compression and expansion.
7. Dolby and dbx are the two most popular noise-reduction systems.
8. Dolby has three models: A, B, and C. Each splits the signal into frequency bands (four bands for A, two each for B and C) and applies compression/expansion.
9. There are two dbx systems (Type I and Type II), which process the entire signal without splitting it into bands.
10. Reverberation is generated either in a reverb chamber or by one of three types of devices: plate reverb, spring reverb, or digital reverb.
11. A *varispeed* makes a tape deck capstan run faster or slower.
12. A *limiter* keeps a signal from exceeding a preset loudness level.
13. A *filter* removes one or more bands of frequencies from a signal.
14. A *pitch-shifter* allows you to vary a tape's speed without affecting the pitch. Conversely, you can raise or lower the pitch without affecting the speed.
15. A *digital delay* duplicates an incoming signal and sends both the original and the duplicate out, but with one delayed behind the other.
16. A *de-esser* is a special filter/limiter, which reduces sibilance.

14

Jingles

A frequently overlooked but nonetheless important aspect of the radio station production manager's job is the creation and implementation of the station's jingle/station identification (ID) package. A station may spend tens of thousands of dollars every 2 or 3 years creating packages of IDs and musical transitional elements designed to keep their call letters in the public's ear and to maintain or create a particular image in the public's mind. Because the station's survival frequently hinges on its public image, especially following a change of format or call letters, the role of the ID package is pivotal.

First, let's define what a jingle is. In the broadest sense, a *jingle* is a short piece of music designed to facilitate and stimulate a listener's retention of a message. A jingle is generally short (:30–:60 and generally much shorter for radio jingles), and the melody is both simple (easy to hum) and catchy—or so the producer and client hope.

Station jingles obviously must have music and lyrics that fit the station. Jingle companies (and good production libraries) can come up with music to fit and supplement any format from classical to country, from all oldies to all talk. Some specific product jingles (like McDonald's "You deserve a break today," for example) have melodies that are not only catchy but are also generic enough to be sung in virtually any style—pop, country, jazz, funk, whatever.

The reason that jingles (whether for radio stations, products, political candidates, state and national tourism departments, whatever) work hinges on two principles. First, I'm convinced that there's a part of the brain that has an affinity for and an amazing capacity to retain melody. How often has a song come on the radio that you haven't heard for 10 or 15 or 20 years, and yet you're able to hum along. You'll probably only remember snatches of the lyrics, but your memory of the melody will be remarkably intact.

Second, because of the brain's affinity for melody, learning can be enhanced by coupling information with music. When you teach a child the alphabet, do you say "This is an A, this a B, this is a C . . . ?" Of course not. You sing the alphabet song. My son was able to sing the alphabet before his first birthday. He had only a vague idea of what he was singing about, but the song stuck.

He learned all sorts of things that we set to music—his address, the spelling of his name, etc. There was an episode of "Cheers" when Coach was helping Sam prepare for a night school geography test by setting the material to the tune of 'When the Saints Go Marching In'" (Al-ba-ni-aaaaa, Al-ba-ni-aaaaa , lo-ca-ted on the A-dri-atic). Funny? Yes, but also effective.

The point of all this is that music aids and often ensures retention. Before the advent of radio, music was disseminated only through sheet music, live performance, and records. With radio instantly putting music into homes nationwide, and with advertisers doing the same with their messages, it wasn't long before the two were combined, and the jingle was born. Although product jingles play a part in the radio producer's job, the radio station's own jingle (or ID) package is the focus of this chapter.

History

Oddly enough, although commercial jingles were commonplace during the 1930s and 1940s, it wasn't until the early 1950s that radio stations realized that the jingle that helped sell a client's product could also promote the station. No one really knows for sure where or when or for whom the first radio ID was recorded, but legend has it that it was produced by Tom Merryman, the founder of TM Productions, for Gordon McLendon's flagship station, KLIF in Dallas, in 1954. This represented a major step up the sophistication ladder for McLendon, one of the pioneers of modern radio programming and production, who once trained a parrot to perch behind him in the control room and shriek the station's call letters.

Whimsy aside, McLendon and men like Todd Storz, Bill Stewart, and Chuck Blore were responsible for keeping radio alive during the years following the advent of TV by leaving the protective womb of the big networks, and by giving their stations a local focus. Formats were devised featuring specific types of music (beautiful music, middle-of-the-road, rock and roll). McLendon also introduced the first all-news and Beautiful Music stations. With the rise of format radio and the perpetual-motion style of the average disc jockey, the station jingles became central to programming strategy. (Although many authorities have anointed the period from the 1930s through the 1940s as the Golden Age of Radio, others of us have a particular affinity for that exciting, unpredictable and dazzlingly creative period from 1950–1965, when format radio and the rise of rock and roll made the radio the best show in town. (The best book that I've found on this subject is Peter Fornatale's, *Radio in the Television Age.*)

During the 1950s and 1960s, Dallas became the jingle mecca of the country, but by the 1970s, producers had sprung up from coast to coast, eager to cash in on the burgeoning, lucrative market. As the number of station licenses multiplied, so did the demand for IDs.

Structure

Jingles differ in form, length, tempo, and content. Here are the most widely used variations.

Form

The various forms of jingles include full sing, donut, end sing, a cappella, and shout.

Full Sing

When the lyric completely fills the instrumental bed, the jingle form is termed *full sing*.

Donut

When the lyric occupies the beginning and end of the instrumental bed, leaving a space in the middle for the local producer to add whatever is needed, this jingle form is the *donut*. Advertisers, especially (but not exclusively) those in less affluent markets, use the professional quality of :30 and :60 donuts to give their commercials credibility and sophistication. As a programming tool, a donut can be used as a place to drop in a short bit of information, such as the time, the temperature, or the announcer's name.

End Sing

The *end sing jingle* features an instrumental bed with no lyrics until the conclusion. The announcer can do a number of things with this type of jingle. Depending on the length of the bed, the announcer can read the weather, talk about what's coming later in the program, or just chatter. Rather than do any of these things in the clear (with no music playing in the background), the music bed, as you'll remember from the chapter on mixing, supports and supplements the voice. When the announcer is done, the end sing jingle jumps from the background, probably with the station's call letters, to be followed by whatever item of business the announcer has planned. The jingle adds continuity to the proceedings, as long as there's no dead air between the three elements—the announcer, the end sing lyric, and the subsequent programming event.

A Cappella

When a jingle has no instrumental bed—voices only—it is *a cappella*. Compared to other forms, a cappellas are rather dull and lifeless, and so most stations don't bother with them. This is a mistake, because, as is shown later in this chapter, you can use a cappella jingles to create new IDs that even your package's producer didn't envision.

Shout

The *shout* jingle form is not as common today as in decades past. Call letters, slogans, clever one-liners, and jock names were often shouted. Vocalists didn't really yell—they spoke the material in an energetic unison, which gave the impression of shouting. Shouts are versatile because they aren't sung in a musical key and can therefore be coupled with or dropped into musical jingles.

Length and Tempo

Jingles can be any length and any tempo. However, these three types are categories unto themselves: shotgun jingles, promo songs, and transitional jingles.

Shotgun Jingles

Short, fast jingles, sometimes only 1–2 seconds long, these *shotgun jingles* are designed simply to punch the call-letters between records.

Promo Songs

Full-length, fully produced *promo songs* laud the virtues of your city or region and of your station. No matter how well produced promo songs are, because of their length, they run a high risk of *burnout* (i.e., if the song's played too frequently, the audience gets sick of it). So, whereas you might get months of use from a small jingle, the useful life of a promo song is considerably less, sometimes only a week or two, depending on the frequency of play and its overall musical and production merit. Promo songs are especially effective when used to publicize a really unusual event, such as a change in format or call letters. Some companies will not only produce a promo song for you, but will also chop the song up into bits and pieces which can be used as individual jingles.

Transitional Jingles

Although most jingles maintain a single tempo throughout, a *transitional jingle* begins in one tempo and ends in another and is used to smooth the transition between two records that have differing tempos (or tempi, if you're a purist). If an announcer's playing a particularly raucous record and wants to segue into a slow, mellow ballad, a fast-to-slow transitional jingle can be inserted between the records, to minimize the negative effect of the extreme shift in tempo.

Content

When Don Imus left WABC in New York, he had a jingle produced for his last day, which featured the lyric "You're fired. . . from WABC." More often a jingle's content runs along one of these lines: station ID, slogan, image, announcer ID, daypart ID, or community ID.

Station ID

The station ID jingle features a station's call letters and sometimes its location.

Slogan

Catchy (the program director hopes) *slogans* abound. A station might want you to remember it as "the music connection," "your station for news," or "the place to be in *[your area]*," During the heyday of the disc jockey in the early and mid-1960s, there wasn't a market in the country that didn't have at least one station claiming to be "the home of the Good Guys."

Image

The image jingle is similar to a slogan but relies more on an emotional approach to elicit a positive response from the listener. A station may want to portray itself as "the station that cares," or "the station that keeps you company," or "the station where your friends are," or simply "your station."

Announcer ID

There isn't a jock alive who doesn't love the sound of his or her own name being sung or shouted. You can have a jock ID sung, with a separate version for each member of the on-air staff. I also like to have the singers produce a shout of each name. Because a shout isn't in a set key, I can drop it into any other jingle I choose, as long as there's a space in the jingle long enough to accommodate the shout.

Daypart ID

Unlike most jingles, which can be played anytime, the *daypart ID* is designed, through specific lyrics or tempo, to be played only at a particular time of the day. It's not unusual to find a jingle package that features a separate jingle for morning, midday, afternoon, and nighttime play.

Community ID
The community ID ties the station to some positive aspect of the community, such as a community's scenic beauty, culture, winning sports team, natural or manmade landmarks, or friendly people.

Assembling the Package

Although attempting to do a job yourself is often laudable, the task of creating good station jingles is best left to the experts. Unless you have access to state-of-the-art equipment and have top-flight arrangers, instrumentalists, and singers at your disposal (not to mention the money necessary to hire them all), don't bother. Listen to demos from some of the major production houses. If you can duplicate what they do with the resources available to you, then go ahead. Otherwise, don't risk wimpy (albeit, well-produced) jingles. To the listener, *your jingles are your station.*

When a large radio station needs an ID package, they get together with the jingle company's creative staff which may include composers, arrangers, and idea people. They then create a package reflecting the station's goals and needs, while staying within the station's budget limitations. However, those limitations better be high, because the tab for the package will run well into the tens of thousands of dollars, depending on the size and complexity of the job. Vocalists and instrumentalists, producers and technicians don't (and shouldn't) come cheap.

Fortunately, that same major-market jingle package can be adapted for use by smaller stations. The singers can be called in to sing *your* call letters, slogan, or image over the existing instrumental tracks. Because you don't need the composer, the arranger, the instrumentalists, or a large production staff, and because you only need to tie up the studio for as long as it takes the singers to do their job, your overall cost is a fraction of what the large station had to pay for virtually the same set of IDs. The only restriction is that your station must be located well away from the original station for whom the package was created. This is only fair. Imagine how the large station would react should they hear the jingles that cost them a fortune being played by a competitor across town who paid substantially less.

Let's say you're the program and/or production manager of a small station in a small- to medium-sized market, and you want to assemble a package. First choose a jingle company whose work you like, and whose sound your listeners won't get tired of. Have the company send you a number of their jingle package demos. Whenever a company creates a package, they also produce a small demo tape of the package, for distribution to people in your position. Call the company, tell them your format or style of music, and have the company send you demos from about 3–5 packages they've done in the past year or two (music styles change, so watch out for old demos), which could fit your format. Listen to the demos, and keep a log of the specific cuts you like. Because your package isn't being made from scratch, you aren't restricted to a single existing package, and you can pick cuts from different packages if you like.

In the beginning you'll probably be dealing with a sales representative of the jingle company. You'll tell him or her the cuts you want, and you'll learn the cost per cut, and how many mixouts you'll get from each cut. Generally, when the company produces a basic jingle for you, they'll also mix 2–4 variations (mixouts) of the jingle for you. For example, a :10 jingle might have a lyric like this:

The station where your friends are,	(:04)
All day and all night long,	(:04)
99 WXXX	(:02)

Although the instrumental bed will not change, dropping out a portion of the vocals could yield a donut:

The station where your friends are,	(:04)
(music bed)	(:04)
99 WXXX	(:02)

It also could yield an end sing:

| (Long bed) | (:08) |
| 99 WXXX | (:02) |

A good package should have at least 10 (preferably at least 12) basic cuts. The cuts should cover a range of lengths and tempos and a mix of vocal arrangements. If male voices dominate your package, or if the instrumental content doesn't vary much, your package will be more of a liability than an asset. Beware of really beautiful or exciting tunes; they sound great when you first hear them, but they burn out quickly, making their usable lifespans short.

When the time comes to talk about how you want the cuts redone to fit your station, ask to speak with the producer or the arranger who'll be in charge of the project. Whereas the musical aspect of your jingles is, to a large extent, out of your hands, the lyrics are your responsibility. Certainly, you'll get help from the producer or arranger: but let's face it: the lyrics are going to have to reflect your station's place in the market. Nobody at the jingle company is going to know your station or your market or your audience as well as you do. Don't relinquish the relyricking job to anyone else. If you're inexperienced and/or intimidated with this aspect of the production, work hand-in-hand with the producer/arranger during the writing phase of the job. Although the salesperson will be happy to help you relyric and plan your mixouts, this is preproduction and not necessarily a salesperson's area of expertise.

Important: Before your call is transferred out of sales, tell the rep that you also want the a cappella cut of each full jingle. The rep may throw the a cappellas in for nothing if your order's big enough, or they may be counted as mixouts. In any event, you want them!

Another aspect of a jingle package is the logo, a short musical phrase incorporated into the cuts, which, over time, becomes something of a signature, like the NBC three-tone logo, which has been the network's trademark for eons. Even though each jingle package has a different logo, pick *one* logo that you like from your demos, and have that logo sung in all your cuts. (Having more than one logo floating around your air would defeat the purpose of a logo: Individuality.) Jingle instrumental beds are frequently flexible enough so that any logo, provided it follows the music's implied harmony, can be sung over the bed. Some radio logos have become famous over the years: WABC (New York), WMAQ (Chicago), and WBZ (Boston) come to mind.

After you've discussed the new lyrics with the arranger, ask her or him to tell you, or to include with the package, the musical keys that the basic cuts are in. Using the compatible keys chart from Chapter 10, you'll not only be able to smoothly segue from jingle into record or from record to jingle to record, but you'll also be able to join an a cappella cut to one of the basic cuts or mixouts in your package (in a similar or compatible key) and thereby create a number of hybrid jingles, especially transitionals. As long as the keys match or are compatible, an a cappella, regardless of tempo, can be joined with a jingle in a different tempo. The hybrid you're producing should be assembled in this fashion:

1. Play the a cappella.
2. Most jingles begin with a brief instrumental pickup, sometimes only a note or two. Make sure this pickup plays *underneath* the end of the a cappella. This overlap is necessary to make the transition between the two elements sound smooth. There must be no break between the a cappella and the jingle.

This assembly is best done on a multitrack recorder. Put the a cappella on one track, and the jingle on another with the two tracks in the sync mode. It may take a couple of attempts to time the entrance of the jingle just right, but it's well worth the effort. Each hybrid you produce is a cut you don't have to pay for. Because individual cuts in your package may run $300 or more, by mixing a dozen or more hybrids, you can substantially increase the value of your investment. Also, with more available jingles to play, the risk of burnout is less.

Don't be intimidated by having to work with music professionals. The jingle business is very competitive, so the people you're likely to deal with will probably have no qualms about going the extra mile to make you happy. I've had arrangers alter melodies, rearrange harmonies, and sometimes restructure an entire jingle. As long as what you want fits the existing instrumental track and the implied harmony, almost anything goes.

Implementing the Package

When you order a jingle package (see Figure 14.1), depending on the company, you'll either buy it outright, or you'll lease it for a set period of time, generally 1–2 years. Always go for at least a 2-year lease, and ask for the option to renew at the end of the lease period at the same (or lower) rate.

Let's say you've ordered a package consisting of 12 basic cuts, and the jingle company gave you 3 mixouts of each cut. In addition, you've mixed 9 hybrids. That gives you a total of 57 cuts. First, cart up the entire package, labeling each cart with the cut's length, tempo, opening (and closing if different) key, and opening and/or closing words. Then divide this stack of carts into three minipackages of 19 carts each. As best you can, try to ensure that these minis are a real mix of cuts—different tempos, styles, lengths, donuts, end sings, and full sings. You're going to put these small packages on the air at 4-month

14.1 Jingle package.

intervals, for example, Group 1 will air from January through April, Group 2 from May through August, and Group 3 from September through December; then restart Group 1 in January, repeating the process for the second year. Assuming judicious jingle play, this staggered rotation minimizes burnout, and allows for full use of your package.

The life expectancy of a jingle is determined by the number of times it's played. If your format calls for 7–8 jingles per hour, you'd better have either a very large package, or the money to afford a new package every year. Novice announcers tend to play lots of jingles, sometimes by choice, and sometimes because management doesn't want the jock to talk, figuring that announcer chatter can be replaced by frequent jingles. ("Just tell 'em the time and temperature, and play the records, kid!") Also, because much of an announcer's life is repetition, there are few events that can set the on-air staff's hearts pumping as quickly as can the purchase of a new jingle package. As a result, it's not unusual to hear a jock run through the entire jingle rack in one show. If you have good, personality-oriented announcers, 3–4 jingles per hour are plenty.

One final word. Your jingles act as salespeople—they help sell your station to your listeners. If there's anything wrong with any of your jingles, call the jingle company and have the problem rectified. The company should correct errors free of charge, especially if they make a mistake like a mispronunciation, for example. Often, when producing jock IDs, a name may be mispronounced. (You can guard against mispronunciation by spelling the names phonetically for the producer.) Don't worry. It happens, and the company will redo the cut. Once, when a new package came into the station, a bunch of us dashed into the production studio to listen to the cuts and almost had a mass coronary! In every cut, the singers had sung the call letters WHEM instead of WHEN! If any aspect of the package is not to your liking, don't put it on the air.

Review

1. A jingle is a short piece of music designed to stimulate a listener's retention of a message.
2. The brain seems to retain a message more effectively when the message is set to music.
3. Jingles were first used to sell products. By the 1950s, radio stations were using them.
4. Jingles vary in form, length, tempo, and content.
5. Smaller stations can take jingles that were produced for big organizations and have the lyrics redone to fit their markets and formats.
6. Choose a good jingle company, and work with the producer/arranger.

A jingle package should contain at least ten basic cuts and as many mixouts as possible.

7. A capella jingles can be mixed with other cuts to form transitional jingles.

8. Avoid having jingles burn out by splitting the package into smaller groups, and rotate the groups of jingles on and off the air every few months.

15

Radio Drama

History

Sadly, radio drama isn't nearly the force it was years ago. Periodically, I hear of isolated groups of devotees attempting to breathe life back into this wonderful form of entertainment with original work. Also, Public Radio, bless 'em, does its best to keep the masterpieces of the past from disappearing altogether. However in the face of stark reality, there's no way enough people are going to be lulled away from television to make radio drama an economically feasible venture. Norman Corwin, the acknowledged master of the form, and probably the finest radio dramatist this country ever produced, was asked on a segment of "CBS Sunday Morning" whether he thought radio drama could make a comeback. His wistful reply was "It will never come back in full force. I'd love to say, 'I think it will.' All I can do is say, 'I hope it will.' But the chances are small."

Technically, any commercial that tells a story can qualify as a radio drama, albeit on a small scale. In this minidrama, a situation is presented, featuring characters who must deal with and attempt to resolve a crisis (through use of some product). If you think about it, everything from *Gone With the Wind* to a commercial for dandruff shampoo fits this formula.

Although the chances are slight that you, as either a professional producer or professional-to-be, will be involved in production of a full-length radio play (outside of a class assignment), there exists no better test of a producer's mettle and all-around skill. Because of its complex and often unwieldy nature, radio drama mandates that the producer develop skills not required by any other facet of the business. Nonetheless, the reward of hearing a well-done, finished production far exceeds the inevitable trials encountered along the way. Radio drama is sound at its most creative and is an exasperating experience for the button-pushing, follow-a-formula producer I hope you never become.

The beauty, indeed the magic, of radio drama, lies in the mind's ability to take aural suggestions and turn them into vivid mental images. In the BBC Sound Effects Library's index is written the following:

The realm of sound is the kingdom of imagination. A car observed is a car; but a car heard—how does it look? Who is in it? Where is it coming from or going to? And that is a very mundane example. No storm or ghost created for a realistic medium like film or stage can match in ferocity or terror the image created by the mind, building on sound.

On the cassette that accompanies this textbook, I've created a storm for you, the kind of storm Arch Oboler, the producer and writer of the acclaimed *Lights Out* radio series would've loved. I simply layered in succession the following sounds, which should be found in any self-respecting radio storm: wind, rain, thunder, the scream of a person in distress, and some moody music. Note how the addition of each sound to the sum of the previous sounds creates a progressively more powerful image. This storm took less than 10 minutes to produce and required four SFX tracks and one music track. I love to play this little opus for snooty TV producers who are convinced of the superiority of their medium, and challenge them to duplicate it on the screen in 10 minutes—or 10 hours.

Besides its obvious ability to entertain, radio drama demonstrated the immense power of sound, the power to influence and motivate an audience to action. This power was demonstrated most notably in the nationwide panic set off by the 1938 Orson Welles/*Mercury Theater on the Air* broadcast of "War of the Worlds", which scared the pants off a large chunk of the American citizenry.

Lest you think that Welles' broadcast was a fluke, an isolated incident of mass hysteria, or an indication that the folks back then weren't quite as sophisticated as we are today, think again. Chuck Blore was the program director at KFWB in Los Angeles when, in order to draw attention to an anti-drug abuse campaign, he had his on-air staff repeatedly break into the regular programming one day with announcements claiming that an amoeba was running amuck in the city. A lot of people apparently didn't know what an amoeba was, and as a result, police and fire control switchboards overloaded as frantic listeners requested information and protection.

In Rochester, New York, a number of years ago, as a practical joke, a disc jockey announced that a large snake had escaped from the zoo and was heading toward a certain area of the city. Instantly, listeners began calling in not only for more information, but also *to report that they had seen the snake!* The power of sound to stimulate the mind is enormous.

Radio drama was a mainstay of radio programming for years and reached its zenith during the decade of 1935-1945. The country's best writers and producers, engineers and effects specialists, musicians and actors all flocked to the microphones and created an art form that was as distinctly American as jazz from New Orleans and musicals on Broadway. Because dozens of people, technical and creative, might be involved in a single production, the cost was often (though not always) exorbitant. Fortunately, though, there were always sponsors available to pick up the tab for a good show in exchange for commercial time.

Then came TV: as the new medium grew in popularity and the sponsors and the dollars changed sides, the radio networks had to sink or swim on their own. They sank. Without the money, the talent could not be hired, and without the talent, live radio drama began an inexorable slide into oblivion. By 1953, radio drama was virtually extinct.

In 1974, CBS tried to resurrect the form with *The CBS Mystery Theater*, an anthology of hour-long original radio plays, with production supervised by one of radio's legendary producers, Himan Brown. The show was fed to CBS affiliates nationwide, who decided whether to air the show. The *Mystery Theater* was a good attempt at artistic, often intellectual fare, but, according to Brown, although the public bought it, the sponsors didn't. *Mystery Theater* was on the air until 1983, a 9 year run. For the first 6 years, there was a broadcast every night. For the last 3 years, performances were cut to five per week. The actors involved were generally unknown, except for the narrator, E. G. Marshall. The production, while technically correct, often consisted of dialogue or monologue, with minimal use of sound effects and minimal use of music for transitions or atmosphere.

According to Himan Brown, it was politics and the economics of contemporary broadcasting that caused CBS to pull the plug. The CBS sales force was more inclined to sell individual spot announcements than to try to pitch a full-hour program to a client. Brown is convinced that there's still a desire on the part of the American public for radio plays, and he tells me that in England, the BBC broadcasts 800 hours of radio plays every year. In 1989, he tried to get PBS interested in doing some original radio theater, in addition to or in place of the classic shows they broadcast, but to no avail. Also, with the zeal of an evangelist, he carries his cause to college campuses and audiences around the nation, determined to bring good radio drama back to the American airwaves.

Also in the early 1980s, the *Sears Radio Theater* went on the air. It featured set themes for specific days of the week, such as westerns on Mondays, mysteries on Tuesdays, and so on. Each day had a regular host, a star such as Richard Widmark or Vincent Price. Also, the overall production was slicker, if less highbrow, than *Mystery Theater*. Nevertheless, Sears's effort met the same fate as CBS's, and by the mid-1980s, radio drama was again only a memory.

Producing a Radio Play

Like a television program, a radio play can be performed live or can be prerecorded. A live performance, while testing your courage and possessing spontaneity not found in a prerecorded production, doesn't allow you fully to show your production expertise, which can only be demonstrated by work in the studio. I have my students produce a play every semester specifically because it's the only exercise I know of that requires them to utilize *all* the production skills and techniques they've been called upon to master—proper equipment usage, mixing, special effects, varied use of the voice, processing—literally everything they know. Not only that, but during play production they'll often run into situations that may not have been addressed directly in class, but that call upon the producer(s) to synthesize solutions from existing knowledge. In other words, they have to be creative.

I'm going to take you step by step through the process I follow with my students in producing a radio play. I warn you that this project will consume a lot of time and will generate a lot of sweat. So does anything of consequence.

The Script

The first step toward producing a radio play is getting hold of a good script. There are two sources: (1) use an existing script, and (2) write your own.

Using an Existing Script

If you have access to any good library, you'll probably find a lot of material written expressly for radio, such as award-winning radio plays, collected plays by radio dramatists, classic radio shows, and stage plays that can be adapted for radio.

During the heyday of radio drama, there were awards presented annually for outstanding radio plays. Many of these collections are still available. They're obviously dated and often trite, but a little rewriting can bring an old chestnut up to date—if that's what you want.

You might also look for collections of plays by the great radio dramatists, especially Norman Corwin. There are a number of collections of Corwin's plays around (as well as at least one biography). Even if you decide not to produce any of Corwin's plays, at least read them. They're short and splendid.

Another option is to transcribe any of the tapes on the market that feature broadcasts of some of the classic radio shows. Episodes of "The Shadow," "The Lone Ranger," "Lights Out," "Gangbusters," and "Gunsmoke," as well as comedies, soap operas, and other dramatic shows are available for purchase, frequently at better bookstores. Also, check your library's audiovisual (AV) section. Remember, however, that this material is copyrighted (and in the case of some tapes, bootlegged), and you're not allowed to make any money for anyone through its use.

Finally, if you can't find anything you like from among the works written specifically for radio, look into plays written for the stage, especially shorter

(1–2 acts) works. If you're looking through stage plays, make sure you pick one that would lend itself to aural presentation. Can the action be re-created on the air? Is there opportunity for use of music and SFX? Will the actors be able to communicate using nothing but their voices? Don't forget that a radio play doesn't have to be serious. Comedies, horror stories, mysteries, period pieces, and romances were all staples of radio's Golden Age.

Writing Your Own

Either make up an original story, or take an existing story, play, or poem, and adapt it for radio. Bear in mind that a radio play is meant to be heard, not read. So as you write, read your work out loud. Remember that subtle nuances, as well as complicated convolutions of language and plot, which may be enjoyed by a reader who can stop and reread a passage may be lost on listeners who cannot mull over what they've heard without losing subsequent information. If you're writing or adapting a script, keep things simple. Activity is important, whether verbal or physical, so avoid long philosophical monologues or soliloquies. That's not to say that your play shouldn't try to present or put forth a moral or a central theme. Present your ideas to the audience through your plot and through your characters, but don't let your script degenerate into a sermon.

Preproduction

With your script in hand, it's time to begin planning the production. This is done outside the studio. Assuming that you're the producer, the first thing to do is to assemble your team and assign acting roles and production duties. The group already should have read the script and decided which speaking part (or parts) they want to play. If someone, for whatever reason, doesn't want to read for a part, he or she will be part of the production team. Everyone should be prepared to work in front of and behind the mike, but if you run across an individual who has an aversion to acting, be sensitive and sensible.

Read through the play. If the voice assignments are working, fine. If not, don't hesitate to swap roles in mid-scene. Let your ear be guided by common sense. The actors must have a clear pictures of the kind of characters they're portraying. They should know how to use their voices (Chapter 8) for maximum effect, and they should be willing and able to let themselves go, even at the risk of feeling silly. Remember that each person should try to sound like he or she is playing a role, not reading a script. Don't read. Act! That means having enough familiarity with the particular part *and the play as a whole* to feel comfortable.

With the various roles assigned, turn next to SFX. Either have one or two people collect the SFX or, if you want to really take charge, you can prepare a list of all the SFX the production will require, and have your SFX people collect them. In either case, you first need to determine what SFX are required. Read slowly through the script and make notations at every place a nonvocal sound could be used. Don't overlook the trivial or the obvious. It's the little sounds that'll breathe life into the play—sounds like footsteps, a door closing, a telephone being hung up. Also, don't make your SFX list based on what you have on hand. Determine what you *need*, and then be prepared to create or improvise the sounds you need but don't have in your library. Remember that SFX make suggestions to the mind, and as such, they don't have to be authentic representations. The early SFX technicians crumpled a ball of cellophane to simulate the sound of fire. If you were to hear crumpling cellophane and real fire, you'd probably have no difficulty telling which was which. However, when you hear the crumpling cellophane at the same time a character screams "Fire!" the sound effect is close enough to the real thing so that your mind, acting on the words being spoken, forces you to hear the sound effect as fire.

Have all the SFX collected onto a single reel of tape. Either have a slate precede each recorded effect or keep a log of what's on the reel. A *slate* is a simple phrase telling what's next on the reel, such as "Take 2," or "Act 1, page 3." You don't have to record them in the order that they'll eventually appear on the final recording. However, if you decide to record SFX out of order, you *must* keep a log.

Next, tackle the music. The music cuts are tougher to collect because music is subjective. A door slam is a door slam, but the suitability of a piece of music depends on how it strikes your fancy at that particular moment. The music serves two functions in your play:

1. It establishes and reinforces a mood.
2. It smooths transitions from scene to scene.

There's no need to have music playing continuously throughout the play. In fact, a lot of short music cues are more effective than wall-to-wall music. Early radio plays, like early talking movies, often had music playing constantly, a carry-over from the days of silent movies. Today, the feeling is that the music should be incidental and should be used to suggest such things like time change, place change, or tension building toward a climax. Also, the music should be fairly nondescript—it should make you feel something, not want to hum along or tap your toe. You don't want to draw attention away from the actors. As has been mentioned in an earlier chapter, movie soundtrack recordings are perfect for your use, as long as you don't violate the copyright restrictions.

As with the SFX, keep all your music cuts on a single reel with the individual cuts slated and/or on a written log. You're going to have to use your best judgment in determining how long each cut of music should be, so make sure you give yourself plenty. You can always fade the music out if it's too long, but if it's too short, you're stuck.

Finally, make sure that these elements (music and SFX) are recorded on the widest tracks and at the fastest speed possible. You're going to have to dub them onto tracks in sync with the voices, and from there onto a master recording. Because you're going to lose two generations of quality, the originals must be of the highest fidelity.

With the music and SFX out of the way, next, determine your stereo imaging. If your production is to be in mono, this phase of preproduction won't concern you. If at all possible bring your play to life through stereo. Read through the script and determine where the voices and/or sounds in each scene will be positioned. Above all, pay great attention to left–right incongruities. Don't let the audience hear a door knock on the left side, and hear the door open on the right. Don't let us hear the character's voice on the left and the character's footsteps simultaneously on the right. Don't let us hear two characters conversing, one left and the other right, but with the ambient noise or background music on only one side. Your stereo placement *must make sense.* You must be able to mentally visualize a scene, and it must be logical. Even though you're dealing with sound and don't have many of the limitations of a visual medium, the same principles of nature and logic must apply.

An important part of this phase of production is ascertaining whether your equipment is capable of delivering the sound you want. Don't sell your gear short, but be honest. If you have a four-track deck, and you have three voices, one left, one right, and one in the center, you're going to need

1. one track for the left voice
2. one track for the right voice
3. one or two tracks for the center voice (If you have a mult, you can record the voice on a single track and split the track via the mult, sending one signal left and the other right for a center image. If you don't have a mult, you'll have to record the voice simultaneously onto two tracks, and throw one left and the other right.)

If, in addition to these three voices, you wanted to add music and SFX, you might find yourself out of available tracks. A bounce to consolidate tracks would be a good idea, or you could mix the voices onto another tape while adding music and SFX directly from disc—a risky, but possible solution. In any event, though it pains me to say this, make sure your equipment is capable of meeting your expectations.

When acting parts and responsibilities have been assigned, when SFX and music cuts have been determined and collected, when imaging has been decided upon, you're ready to go into production.

Production

I'm going to show you how to produce a play using the following studio gear:

1. a four-track deck for recording raw tracks
2. a two-track deck for mixing your master production
3. a board with four pots (excluding mike and turntable pots) for mixing

This isn't a lot of equipment, but it will suffice. If you have more, the process will be easier. If you have less, don't give up. Just be sure, before you begin production, that you have enough equipment to do the job.

Your first item of business is to lay down the voice tracks. If the entire cast has lots of free time and can meet frequently, fine. More often, though, the production will be at the mercy of individual schedules, so be prepared to record the voice tracks out of order. If some of the actors have particularly tough schedules, make every attempt to record them first; otherwise, you may find that these people will inadvertently hold your production hostage.

If two or more voices share a scene, you have to decide whether to mike the voices individually or to have them share a mike. I favor separate miking because it allows for greater freedom in mixing, although if the actors' voices seem evenly matched in terms of power, I may have them share a single mike. If the voices are to be positioned in different parts of the aural field, then there's no choice. You must use separate mikes. When using separate mikes I like to keep the actors well apart. If I don't have much room, I'll position the actors so that they either face away from each other, or face toward a nonreceptive part of the other actor's mike. The possibility of *crosstalk* (one person's voice being picked up by the other person's mike) will be lessened, as well as the possibility of wave cancellation. (Wave cancellation was discussed in Chapter 11, as was flanging. In wave cancellation, if two identical tones are combined, but one is half a cycle behind the other, the crest of one wave will occur simultaneously with the trough of the other, and vice versa, thus causing the two waves to cancel each other. If two mikes are being fed into a single channel, and if the mikes aren't placed correctly, an actor's voice may be picked up not only on his or her own mike, but also by the mike a distance away. Even though the mikes may not be very far apart, the sound hitting the distant mike will be slightly behind the sound striking the close mike. Should that delay constitute half a wavelength, or an odd-numbered multiple of the half wave, then some portions of the sound will cancel each other out. Because the sound reaching the distant mike will probably be substantially lower in level, there's no chance that sound will totally disappear, although it will sound strange and will be lower in level than it should be. If the actors' mikes are being fed into different channels, wave cancellation is not a factor.)

15.1 Multiple mike mixer. (Courtesy of Shure Brothers Inc.)

Many smaller studios utilize a multiple mike mixer (see Figure 15.1). Signals from two or more mikes are sent into the mixer, where their levels are individually controlled. The signals are then combined into a single signal and sent to a common pot on the board. Even though the original mike signals are controlled separately, they do mix when they're sent to the board, so beware of cancellation. If you're using a mixer of this sort, first set the mixer's master level control in the middle of its range. Then have the actor with the quietest voice get on mike, and set this person's level. If the voice is so quiet, and the gain for that mike is cranked up so far that you start hearing noise, back the mike gain off, and boost the master a bit. Once the quiet voice is set, the others shouldn't present a problem.

It's better to have a little noise spread to all the tracks than to have a single very noisy track. If the signal from the mixer is noisy, before the signal reaches the board you should route it into an EQ and slightly lower the frequencies around 8 kHz.

Don't be afraid to improvise. For example, if you need a milling crowd but you only have a couple of voices, record the actors as they walk around the mike mumbling. Rewind the tape, and then either record them again in sync with the first track, or bounce that first track onto another track in the play mode (slap-back echo), at the slowest speed possible. The slower the tape speed, the greater the delay, and in this case, you want a hefty delay to make the overall sound more chaotic. If the echo is far enough behind the original sound, the ear will hear the echo as a distinct sound, and not as an echo. I suppose this could be considered a corollary of the Haas effect.

Regardless of the number of voices being recorded at a given time or the number of mikes in use, assuming that you're planning some stereo imaging later on, the voices should eventually occupy only two tracks on your tape, one to be thrown to the left and the other to the right. If you have an eight-track (or more) tape deck, put each voice (assuming you have six or less at any given time) on a separate track, and then bounce these tracks down to two.

If you have a four-track to work with, you need to record the voices directly onto only two tracks. This procedure, of course, calls for the voices to be perfectly balanced and mixed *as they are being recorded.* If you have two four-track decks, record the raw voice tracks on one machine, and then bounce the tracks onto two tracks of the other machine. Also make sure you leave space (15 seconds will do) between scenes on the voice reel for transitional music, which will be added later.

Working with amateur actors can either be a source of excitement and fulfillment or mental and emotional derangement. Realize the limitations of your actors, but don't be straightjacketed by them. Encourage the players to speak their lines instead of reading them. There's a difference, and preparation is the key. If your players are well rehearsed, they'll be able to concentrate on their delivery, not merely on avoiding tripping over words.

A lot of people have a secret desire to act but hesitate to do so because of fear—fear of failure, fear of an audience, fear of inadequacy, or fear of appearing silly. Because with radio there's no visible audience present, it's not uncommon for introverted amateurs to blossom on mike. I know a lot of fine radio announcers who are so shy and introverted that they have difficulty functioning in a social setting, but when they're behind a mike, they know the audience can't see them and that the audience can be made to disappear with the flick of a switch. This power of anonymity often gives people the confidence to open up and to pretend to be someone or something else—friendly, funny, outgoing, or sexy. If you could see what your favorite radio personalities really look like, you'd probably be shocked. For some reason, we tend to imbue the individual behind a disembodied voice with good looks and a winning personality.

As with music and SFX, if you're recording the scenes out of order, slate the takes as you put the voice tracks on a large reel, and keep a running log. If corrections have to be made at a later time, it's always best to bring all the actors involved back into the studio, and to redo the scene. If the problem is with only a single character, and you've recorded all the actors on separate tracks, the next best option would be to bring the lone actor back and rerecord only his or her lines. Instead of carrying on a conversation with other actors, the person called back will monitor the other actor's lines (in the sync mode), and rerecord the problem track.

When this phase of the project is finished, you should have three reels in hand, one with the voice tracks, one with all your SFX, and one with all your music cuts. Now it's time to sew all the pieces of the quilt together.

Postproduction

The final phase of production is not only the most tedious, but also the most crucial. Mistakes made here are often irreversible and tough to camouflage, so be slow, methodical, and ultracritical. Don't let anything that isn't exactly as it should be get by you. Working alongside someone is a help here—what one person might miss, the other could catch. This is not the place for an inflated ego. The best producers in the history of radio assembled teams to work with them in all phases of production. Why should you be any different?

Assuming you're working with a four-track deck, and you've combined your voice tracks onto two tape tracks (one going left and one going right), you're going to dub the contents of your SFX reel onto one of the two remaining tracks, and the music tracks onto the other. If the voices are on tracks 1 and 3, you might want to put the SFX on 2 and the music on 4. Remember to dub with the four-track in the sync mode, and remember to put the SFX and music onto the four-track tape at full volume—0 Vu.

In all of this, your timing is crucial. Timing your SFX to your voice tracks is easy, but timing the music cuts requires a little finesse. You'll use music *between all* your scenes to suggest passage of time or change in place, and *during some* scenes to reinforce or heighten the mood. I don't usually like to use music during a scene unless the scene's action is building in intensity toward an emotional peak and the addition of music will reinforce this progression. If your play has a structure similar to a short story (and most short radio plays do), all the action will be directed toward a single climax. Putting appropriate dramatic music under a scene as the action builds inexorably to a peak is always a good idea.

Between scenes, the music should start almost immediately after the last word in a scene is spoken. A delay of ½ to ¾ of a second is plenty. If the delay is shorter, the music will draw attention away from and thereby lessen the impact of a scene's final words. If the delay is too long, the music will be perceived only as part of the following scene, rather than as a bridge between two scenes.

The delay between the start of a music transition and the start of the next scene will be longer, sometimes up to 10 seconds, depending on what your ear likes. Again, if the delay is too short, the illusion of time and/or place change won't be adequately communicated. If the music's too long, the audience won't know what's happening. (Is the play over? Is there more?) Here's how to make your music transitions effective:

1. If the music is to be a bridge between, say, Scene 2 and Scene 3, cue up Scene 3 on the voice reel and the desired music track on the music reel.

2. Hearing the last few words of Scene 2 in your head, roll the music. You're not recording anything yet—you're just listening.

3. After 8–10 seconds, simultaneously roll Scene 3 and smoothly fade out the music under the voice (or whatever the scene begins with), so that there's a little overlap (1–2 seconds). Don't let the music keep trailing on and on.

4. If the timing between the music and the start of Scene 3 sounds okay (let's say the music played for 8 seconds before the scene started), recue the voice track and then back the tape 8 seconds further using a timer, if the deck has one, or using the backtiming technique from Chapter 11. Put the deck into the sync mode.

5. Recue the music tape, and dub the music onto the backtimed voice track tape.

6. Dub Scene 2 onto a master tape.

7. Cue Scene 3 at the start of its 8-second music lead-in, and dub onto the master tape a few seconds (4 or 5) after the end of Scene 2.

8. The space between the end of Scene 2 and the start of the transitional music was made deliberately long, so that you'd be able to edit it down to size (less than 1 second).

9. Play the entire transition (end of Scene 2, music, start of Scene 3), and listen to how it sounds. It should be fine.

If you're nervy and want to save editing time, try this after Step 6:

7a. Cue Scene 3 to the start of its 8-second lead-in.

8a. Back the master tape 5–10 seconds from the end of Scene 2, and roll the master tape in the sync mode only. There's no recording happening yet.

9a. When Scene 2 (on the master tape) ends, you're going to wait a fraction of a second (or whatever will sound best) and roll the music at the head of Scene 3. Does the transition sound okay? If not, rewind the tapes, and repeat the process, adjusting the delay interval between the end of Scene 2 and the start of the music.

10a. When you've decided on the timing, rewind the tapes one more time, but this time, after Scene 2 ends, keep the master tape rolling. A split-second before you hit the music on the other deck, hit the master deck's record button.

If your timing was right, no editing is necessary. The music will kick in following the end of Scene 2, and after 8 seconds, Scene 3 will begin. If you don't have quick reflexes and steady nerves, don't try this technique! You'll risk hitting the record button too soon, thereby recording over the tail of Scene

2. Also, listen carefully to see whether hitting the record button put a pop or a click onto the master. If so, forget this procedure, and stick with the previous technique, which requires editing.

Assemble all your musical transitions this way—that is, backtiming the music to the head of the subsequent (rather than following the tail of the previous) scene. Record each completed scene onto the master tape in the correct sequence. After a scene is dubbed onto the master, check to make sure the level is consistent with the previous scene. If so, move to the next scene. If not, redub the scene, and alter the levels accordingly. Now's the time for any changes, when the scene you're working on is, for the moment, the last thing on the master.

Let's say you dub Scene 3 onto the master, and an imperfection in the scene slips by you. You continue and dub Scene 4. Should you discover the goof in Scene 3 later, going back to fix it will be very difficult, if not impossible, excluding the possibility of redubbing everything from Scene 3 to the end. If you want to spare yourself a lot of grief, remember:

Listen *critically* to *everything* that goes onto the master.

I know for a fact that producers are responsible for a disproportionate chunk of the coffee, cigarette, and No-Doz consumption in this country. Be aware of your limits. If you've been laboring in the studio for 6 hours straight, there's no way you're going to be sharp. also, it's when you're a little bleary, in that fuzzy Twilight Zone between alertness and exhaustion, that mistakes can slip by. Worse still, you'll hear the mistakes and let them go, figuring that no one will notice them. There's nothing wimpy about taking an occasional break to maintain your acuity. Know when to go home.

When your play is completely mastered, sit back and enjoy it. Even though you probably know the production inside and out, listen as if you're hearing it for the first time. Let yourself be sucked into the sound. This is how it was, when in the not-too-distant past, radio and our imagination joined to make magic. You may not know it, but you've helped keep an art form alive. Those of us who love it thank you.

Review

1. Producing a radio drama is the best all-around test of a producer's skill.
2. Radio drama works because the mind of the listener automatically creates vivid mental images when prompted by aural suggestions.
3. Radio drama hit its zenith between 1935 and 1945, notably in the work of Norman Corwin.

4. In the 1970s and 1980s, unsuccessful attempts were made to put radio drama back on the air.
5. A play can be performed live or preproduced on tape. The live performance has more spontaneity, but the prerecorded drama allows for greater experimentation and more polish.
6. Award winning radio scripts are available in local libraries. Classic radio plays are available on audiocassettes and be transcribed.
7. If you write your own play, avoid long monologues, which deaden the action.
8. If you manufacture your own sound effects, you don't have to duplicate the real sounds—just come close. The listener's mind will make it realistic for you.
9. Music in the drama either reinforces a mood or smooths transitions from scene to scene.
10. Use music only when needed. Avoid wall-to-wall music.
11. When planning your stereo imaging, be aware of left–right incongruities.
12. If actors share a common mike, position them to ensure that their voices are balanced. If the actors are on separate mikes, listen for wave cancellation.
13. Although the production should be planned before entering the studio, don't hesitate to improvise.
14. During postproduction, don't let imperfections slide by. They'll be tough to repair later on. Listen critically to everything that goes onto your master.

Afterword

I remember stepping out of a roller coaster at an amusement park in Massachusetts when I was a kid. The attendant, noticing my chalky complexion and terrorized expression, greeted me with a cheery "How'd you like the ride?" I didn't know whether he was being serious or sarcastic. I still don't.

Now that you've finished *Creative Radio Production* I feel like asking you the same question.

And I'm very serious.

My goal with this book was not to make you into a producer, but to give you enough knowledge, encouragement, and guidance so that you could make yourself into a producer. Producers are made, not born, and they are not made in classrooms or between the covers of books. They're made in studios.

With the completion of this book you're ready to take off on your own. And, with a little luck, the learning process you'll begin now will never end. There are no experts in this field, though some folks certainly know a lot. An *expert*, according to Harry Truman, is someone who's afraid to learn anything new, because then he might not be an expert anymore. Every semester I learn something new from students who have flashes of inspiration. And I'm not talking about new technology; I'm referring to finding new ways to do standard stuff: Creativity.

That's why I love production so much. It's protean. It's always in a state of flux. My colleagues ask if I get bored teaching the same subject every semester, and I tell them in complete honesty, "How could I be bored? The students take the information and do unique things with it all the time. Some of the work shows minimal creativity (not willing to take chances, playing it safe), and some is absolutely dazzling."

Early in my career, when someone at a social gathering would ask the nature of my work, I'd spout off a rather wordy and somewhat technical (might as well try to impress) synopsis of my job responsibilities.

Today, being older and (I hope) wiser, I put the response more succinctly. "I play with sound."

Since the variety of sound is infinite, and since the creativity you use to fashion sound in your mind's ear is similarly limitless, how could boredom or repetition be a factor? Remember the closet I referred to in the Preface? Prop that door wide open.

Is the ocean boring? Or a Beethoven symphony? Or Michelangelo's marble "Pietà"?

Or a roller coaster?

Hope you liked the ride.

Appendix 1: *No Exit*, an Original Radio Play

Here is a sample radio play for you to try. Depending on what you do with it, the finished product should run somewhere between 12 and 15 minutes. There are two major and three minor characters, as well as ample opportunity for SFX and music. I've intentionally omitted suggestions for SFX, to avoid influencing your decisions. I've also indicated places for musical bridges and scene changes but without implying any specific type of music. You decide. It's up to you to dress the set, draw the characters, and bring the story to life. Shakespeare did the same thing. Look at any of his plays. He has only two stage directions: enter so-and-so, and exit so-and-so. The rest was up to the director.

No Exit

GEORGE: Helen, help me with the door.
HELEN: I'm coming. Why can't you open the door yourself?
                   (Door opens)
   George, what is this?
GEORGE: It's what we've wanted but could never afford. Our own computer. Help me with this, will you?
HELEN: But you know we decided we couldn't get one until . . .
GEORGE: Until we had more money. I know. But I couldn't pass this deal up. Isn't it a beauty?
HELEN: It is nice. But where did you get it?
GEORGE: Well, I was on my way home . . .
HELEN: Late, as usual.
GEORGE: It's not my fault. Henderson gave me eight more files to work on, and it was either work late tonight or come in on Saturday.
HELEN: Why do you let him walk all over you like that?
GEORGE: Because he's my boss. What else can I do?
HELEN: You could try standing up to him for once.

GEORGE: Listen, we've been through all this before. I know he treats me like dirt. But it's a good company. And unless he dies or disappears, I've got no choice but to put up with him.

HELEN: All right, let's change the subject. You were on your way home.

GEORGE: Right! And I passed by the police station, and they were having some kind of an auction. Stuff they confiscated from drug dealers and people like that.

HELEN: You mean this belonged to some psycho who might decide he wants his computer back and track down the fools who bought it?

GEORGE: No chance. This actually belonged to some bank teller who decided to permanently borrow some of the bank's funds, probably for an extended stay in another hemisphere.

HELEN: So where is he?

GEORGE: They don't know.

HELEN: Who is he?

GEORGE: I didn't ask. Listen, the guy disappeared without a trace; and they don't expect him to show up again—ever.

HELEN: And the Titanic didn't expect to hit an iceberg.

GEORGE: Very funny. But look at this thing. Color monitor, dual disk drive, hard disk with tons of memory. . . . It'll do everything we want. Here's the receipt.

HELEN: You only paid $400 for this!

GEORGE: That's all. They started the bidding at $400, and no one else wanted it.

HELEN: Anything else come with it?

GEORGE: A box of disks. The police kept most of them in case there was anything on them they could use to track the guy down. Actually, all they left me were a bunch of games.

HELEN: This may be a silly question . . . but does it work?

GEORGE: Beats me. Let's find out.

(Bridge)

GEORGE: I think that'll do it. Did you plug the monitor in?

HELEN: Aye, aye captain.

GEORGE: OK. Here goes. . . . Hey, it works!

HELEN: For $400, it better work.

GEORGE: Hand me one of those disks.

HELEN: How about this one? *Artifical Intelligence.*

GEORGE: Fine. Let's see what it does . . . There we go!

HELEN: (reading) ENTER YOUR NAME.

GEORGE: G-E-O-R-G-E. And we hit enter.

HELEN: (reading) HELLO, GEORGE. ASK ME ANY QUESTION WHICH CAN BE ANSWERED WITH A YES OR NO RESPONSE.

GEORGE: OK. (typing) Is-my-wife-beautiful?

HELEN: I hope this machine's user-friendly. What's it say?

GEORGE: (reading) THAT IS A DISTINCT POSSIBILITY.

HELEN: Sounds like a cop-out to me.

GEORGE: What'd you expect? It's programmed to give you vague, generic answers. Kind of like the horoscope in the newspaper.

HELEN: This is going to be a barren source of amusement.

GEORGE: Let's try this. (typing) Is-my-boss-a-jerk? (pause)

HELEN: (reading) I AGREE WITH YOUR ASSESSMENT.

GEORGE: Hey this machine's not so dumb after all.

HELEN: Neither am I. Which is why I'm going to bed. And I suggest you do the same.

GEORGE: In a little while. I want to play with this some more.

HELEN: Have a good time. But remember, Henderson expects to see you tomorrow with both eyes open.

GEORGE: I know, I know. Get to bed. I'll be there . . . eventually.

(Bridge)

(tired) I think I've had it for tonight. One more question. (typing) Will-I-sleep-well-tonight? (reading) ONLY IF YOU DON'T DREAM ABOUT YOUR BOSS. . . . Wait a minute. I must be more tired than I thought. Let's try this. (typing) Do-you-know-my-name? (reading) YOUR NAME IS GEORGE. DO YOU KNOW MY NAME? This is crazy. (typing) What-is-your-name? (reading) MY NAME IS ERIC.

HELEN: (from a distance) George, it's 2 A.M. Come to bed.

GEORGE: In a minute. All right, Eric, (typing) Do-you-have-any-advice-for-me? (reading) LISTEN TO YOUR WIFE AND GO TO BED. (typing) Am-I-dreaming? (reading) NO. DO YOU REQUIRE PROOF? (typing) Yes. (reading) ENTER THE NAME OF SOMEONE YOU HATE. That's no problem. (typing) Fred-Henderson. (reading) HIT THE KEY MARKED *SAVE*. Ok . . . (strike key) Save. (pause) Now what do I do? (typing) Now-what-do-I-do? (reading) GO TO BED. GOODNIGHT. Have it your way, Eric. Pleasant dreams. (switch computer off)

(Scene change—Office)

GEORGE: Hi, Sally. Sorry I'm late.

SALLY: No problem. He's not in yet.

GEORGE: I beg your pardon.

SALLY: Mr. Henderson. He's not in yet.

GEORGE: Now there's a first. Fred's never late.

SALLY: He is today.

GEORGE: (concerned) Any idea why? Did he call or anything?

SALLY: Not a word. And anyway, why do you care. He's not exactly your best friend.

GEORGE: I know. It's just that. . . . I mean something's happened that's hard to explain.

SALLY: Like what?

GEORGE: Like forget it. I'm just babbling. Didn't get to sleep 'till late. Anyway, I've got work to do. Let me know if he comes in, will you?

SALLY: Ok.

(Scene change—Home)
(Door closes)

HELEN: Hi, honey.

GEORGE: Hi.

HELEN: What're you so glum about?

GEORGE: Fred Henderson never showed up today.

HELEN: And you're upset because it wasn't declared a national holiday, right?

GEORGE: This isn't funny.

HELEN: Why are you upset?

GEORGE: Because I may have caused something to happen to him.

HELEN: Like what?

GEORGE: I don't know. . . . But the computer might.

HELEN: What?

GEORGE: Helen, you're not going to believe this, but last night, when I was playing with the computer, it started giving me answers to my questions . . . concrete answers, like it was actually communicating with me.

HELEN: You were exhausted, and your mind was playing tricks on you.

GEORGE: No, really. It told me that its name was Eric, and that I should listen to you and go to bed. It asked me for the name of someone I don't like, and I typed in Fred's name, and today, for the first time in history, he doesn't show up for work.

HELEN: Are you saying that the computer made Fred take the day off?

GEORGE: Don't you understand? Fred didn't take the day off. He's disappeared. He's not at home. No one's seen him. No one's heard from him. He's gone.

HELEN: And you think the computer had something to do with it?

GEORGE: I'm about to find out.

(Bridge)

GEORGE: I want you to read what comes up on the screen. Then you tell me if I'm going crazy.

HELEN: But George . . .

GEORGE: Just read. That's all I want you to do. Ready?

HELEN: I suppose.

GEORGE: Here we go. I enter my name G-E-O-R-G-E.

HELEN: (reading) HELLO, GEORGE. I saw this yesterday.

GEORGE: Just wait. (typing) What-is-your-name?

HELEN: (reading) MY NAME IS ERIC.

GEORGE: (typing) Where-is-Fred-Henderson? (pause)

HELEN: Nothing.

GEORGE: What do you mean?

HELEN: Nothing came up. Look for yourself.

GEORGE: Let me enter the question again.

HELEN: But George, doesn't it have to be a yes or no question?

GEORGE: Quiet! (typing) Where's-Fred-Henderson?

HELEN: Here it comes. (pause) This is impossible.

GEORGE: Let me see what it says.

HELEN: It says, HE'S IN HERE WITH ME.

(Scene change—Police Station)

POL: (shuffling papers) That was the personal computer, right?

GEORGE: Right. And the name was George Bradley.

POL: Bradley. Right, here we are. One CO-ROM III computer. . . . you paid
     $400, right?

GEORGE: Right.

POL: What do you want to know, Mr. Bradley?

GEORGE: I'd like to know who owned it before me.

POL: Why?

GEORGE: Well. let's just say there are some glitches I need help working
     out. The previous owner might be able to give me a hand.

POL: According to this, the previous owner's whereabouts are unknown.

GEORGE: Do you have his name?

POL: What does it matter? Didn't you hear me? The guy's gone. He committed
     a federal offense skimming that bank, and if the FBI can't locate him,
     believe me, you can't either.

GEORGE: Humor me. Please. What was his name?

POL: All right, if it's that important. Says here the guy's name is Jason.

GEORGE: Jason what?

POL: No, Jason's his last name. His first name . . . is Eric.

(Scene change—Home)

HELEN: What're you doing?

GEORGE: I'm running the output of the computer through this voice synthesizer I borrowed from my office's computer center. Instead of reading what the computer has to say, we can hear it.

HELEN: George, this really scares me. If you're right, if this Eric Jason or his spirit or something is inside this computer, I think we should destroy it.

GEORGE: Absolutely not! For two reasons. First, I just shelled out $400 for this thing, and I'm not about to eat that kind of a loss. Second, Eric Jason stole over 3 million dollars. Either he's got it in there with him, or he's hidden it somewhere out here. In any case, it can't do him any good where he is. Maybe we can make a deal—his freedom in exchange for a piece of the pie.

HELEN: I won't let you. He's dangerous. Did you forget that somehow he figured a way to hide inside a computer? The man is obviously brilliant.

GEORGE: And I'm not, is that it? I'm not smart enough to match wits with him, is that what you're trying to say?

HELEN: No, George, it's just that . . .

GEORGE: Get out of here, Helen.

HELEN: But George . . .

GEORGE: Now! (Helen leaves—door slams) All right, Eric. Now it's you and me.

(Scene change—street)

HELEN: (talking to herself) I never should have left him alone. He's only thinking of our future. One of these days I'll learn to support him instead of putting him down. Ok, Helen, talk is cheap. It's time to help your husband.

(Scene change—home)
(Helen knocks on the door to the computer room)

HELEN: George? George, can I come in? Honey? (opens door) George? Where are you?

MAN: Perhaps I can help.

HELEN: (gasps) Who are you? How did you get in here?

MAN: Sorry I frightened you.

HELEN: But who are you?

MAN: Just a friend of George's.

HELEN: What're you doing here? And where's George?

MAN: Actually, we made a deal. A very tough bargainer, your husband. At any rate, he got me out of the computer, and I gave him half of the money I borrowed from my bank.

HELEN: Oh, my God. You're . . .
MAN: That's right. I'm Eric Jason.
HELEN: What've you done to my husband?
MAN: Not a thing.
HELEN: Then where is he?
MAN: He's right here. (typing) Say hello to Helen, George.
GEORGE: (from inside the computer) Helen, help me. Help me. Help me.

Appendix 2: Table of Contents to the Cassette Tape

□ □ □
□ □ □
□ □ □

Suggested Bibliography

Alten, Stanley R., *Audio in Media*, 3rd ed., Belmont, CA, Wadsworth Publishing Co., 1990.

Anderton, Craig, *Home Recording for Musicians*, New York, Amsco Publications, 1978.

Backus, John, *The Acoustical Foundations of Music*, 2nd ed., New York, W.W. Norton and Co., 1977.

Bannerman, R. LeRoy, *On a Note of Triumph: Norman Corwin and the Golden Years of Radio*, New York, Carol Publishing Group, 1986.

Book, Albert C., Cary, Norman D., and Tannenbaum, Stanley I., *The Radio and Television Commercial*, Lincolnwood, IL, NTC Business Books, 1988.

Dearling, Robert and Celia, *The Guinness Book of Recorded Sound*, London, Guinness Books, 1984.

Everest, F. Alton, *Handbook of Multichannel Recording*, Blue Ridge Summit, PA, Tab Books, Inc., 1981.

Everest, F. Alton, *The Master Handbook of Acoustics*, Blue Ridge Summit, PA, Tab Books, Inc., 1981.

Fornatale, Peter and Mills, Joshua, *Radio in the Television Age*, New York, The Overlook Press, 1984.

Fox, Ted, *In the Groove: The Stories behind the Great Recordings*, New York, St. Martin's Press, 1986.

Gross, Lynne and Reese, David E., *Radio Production Worktext: Studio and Equipment*, Boston, Focal Press, 1990.

Hall, Claude and Barbara, *This Business of Radio Programming*, New York, Billboard Publications, 1977.

Hilliard, Robert L., *Radio Broadcasting: An Introduction to the Sound Medium*, New York, Longman, Inc., 1985.

Keith, Michael C., *Broadcast Voice Performance*, Boston, Focal Press, 1989.

Keith, Michael C., and Krause, Joseph M., *The Radio Station*, 2nd ed., Boston, Focal Press, 1989.

Maestas, Bobby, *Recording Sessions*, Newbury Park, CA, Alexander Publishing Co., 1989.

Mantell, Harold, ed., *The Complete Guide to the Creation and Use of Sound Effects for Films, TV, and Dramatic Productions*, New York, Films for the Humanities, Inc., 1983.

Martin, George, *All You Need is Ears*, New York, St. Martin's Press, 1979.

McIan, Peter, and Wichman, Larry, *The Musician's Guide to Home Recording*, New York, Simon and Schuster, Inc., 1988.

Milano, Dominic, ed., *Multi-Track Recording*, Milwaukee, WI, Hal Leonard Books, 1988.

Mott, Robert L., *Sound Effects: Radio, TV, and Film*, Boston, Focal Press, 1990.

Oringel, Robert S., *Audio Control Handbook, 6th ed.*, Boston, Focal Press, 1989.

Rapaport, Diane Sward, *How to Make and Sell Your Own Record*, Jerome, Jerome Headlands Press, 1988.

Ries, Al and Trout, Jack, *Positioning: The Battle for Your Mind*, New York, McGraw-Hill Book Company, 1981.

Sterling, Christopher H. and Kittross, John M., *Stay Tuned: A Concise History of American Broadcasting, 2nd ed.*, Belmont, CA, Wadsworth Publishing Co., 1990.

Young, James Webb, *A Technique for Producing Ideas*, Lincolnwood, IL, NTC Business Books, 1988.

Index